THE STUDENT SCIENTIST
EXPLORES ENERGY AND FUELS

THE STUDENT SCIENTIST EXPLORES ENERGY AND FUELS

By
William A. Kaplan
and
Melvyn Lebowitz

RICHARDS ROSEN PRESS, INC.
New York, New York 10010

Published in 1976 by Richards Rosen Press, Inc.
29 East 21st Street, New York, N.Y. 10010

First Edition

Library of Congress Cataloging in Publication Data

Kaplan, William A
  The student scientist explores energy and fuels.

  (Student scientist series)
  1. Power resources. 2. Fuel. I. Lebowitz,
Melvyn, joint author. II. Title.
TJ163.2.K36   333.1   74–34077
ISBN 0–8239–0338–9

Manufactured in the United States of America

To
Arlene and Lori

About the Authors

Dr. William A. Kaplan, a member of the faculty of Brooklyn Technical High School since 1961, has spent most of his professional life teaching courses on the materials of science and industry. While an instructor of industrial processes, he continued his graduate studies and in 1970 received the Doctor of Education degree from Columbia University.

As a member of the Executive Board of the New York Chapter of the American Society for Metals and as chairman of its Student Activities Committee, Dr. Kaplan was responsible for developing and coordinating in-service courses in materials science for high-school science teachers.

Among professional honors received by Dr. Kaplan are the outstanding teacher award of the Chemists Club of New York and the education award of the American Society for Metals. He is a member of the educational honor societies Phi Delta Kappa and Kappa Delta Pi.

Born in New York City, Dr. Kaplan received a Bachelor of Arts degree from Cornell University. After two years of military service in Korea and Japan, he earned the Master of Arts degree in science education at Columbia University Teachers College and began his teaching career. He lives in New York City with his wife, Arlene.

Melvyn Lebowitz, a guidance counselor at Brooklyn Technical High School, is particularly aware of the importance of maintaining the relevance of a school's program of study for modern youth and believes that a knowledge of scientific and technical developments is essential to their growth as citizens.

After having received a Bachelor of Science degree in chemistry at City College in 1961 and having completed several engineering courses, he began his professional career as a chemical engineer. Becoming interested in science education, he returned to City College to pursue that field. He became a teacher of materials at Brooklyn Technical, and after earning a Master's degree in science education, he taught several other subjects, including chemistry and biology. He has also participated in the development of a modern curriculum for the materials science course and has served as assistant chairman of that department.

Mr. Lebowitz lives in Englishtown, New Jersey, with his wife, Lorraine, and their two children, Jonathan and Wendy.

Acknowledgments

The authors wish to express appreciation to the following persons for their assistance:

To Dr. Louis Weiss, principal of Brooklyn Technical High School; to Mr. Herbert Tucker, chairman of the Materials Science Department; to Mr. Max Kohn and Mr. Martin Starfield, former chairmen; and to Mr. William Langweil and Mr. William Clarvit, members of the department, for their encouragement and support of our efforts.

To Mr. Carl T. Goldstein of the Atomic Industrial Forum and Mr. George Millman of the American Electric Power Company for their very constructive suggestions.

Contents

THE STUDENT SCIENTIST
EXPLORES ENERGY AND FUELS

I

The Need for Energy

Energy and power are needed in order to do work. People working by hand use muscle power. Horse-drawn wagons and ox-drawn plows run on the muscle power of animals. In the United States, most of the heavy work is done by the mechanical power of machines. Electrical power is used to light our homes. Automobile engines run on gasoline power, and many truck, bus, and railroad engines are powered by diesel fuel. No matter what its source, the more power we have available, the more work we can do and the better we are able to live. That is why energy is so important to us.

Energy must be available in abundant amounts in order for people to develop a high standard of living. Should energy become scarce for any reason, the way in which we live will be greatly affected. Have you ever experienced a power failure in your area? In 1968, on a November evening, a great electrical power failure occurred in many of the Northeastern states. Farms, towns, and great cities alike were blacked out. The results were catastrophic. Thousands of people were trapped below ground in stalled subway cars in New York City. Trolleys ground to a halt in Boston. Refrigerators and freezers began to defrost, and food started to spoil. In hospitals, surgeons operated on their patients under makeshift emergency power. Street lights and traffic signals stopped working, and automobiles slowed to a crawl as intersections jammed up with vehicles. Many people did not reach home until the following day. Fortunately, that blackout was corrected in most areas within twenty-four hours; but its effects remind us of our everyday reliance upon energy, and how easily we take its availability for granted.

How Have Our Energy Resources Been Developed?

The relative abundance of power and our ability to use it have been developed only in recent times, mainly in the United States, Canada, countries of Europe, and a few other technologically advanced nations. In prehistoric times, human beings were at first limited to their own muscle power. They lived in small groups, gathering edible roots and fruits, hunting animals for meat, and chipping rocks and sticks into crude tools. Living in caves or simple pit houses, these Stone Age people were at the mercy of every storm, drought, freeze, or other natural catastrophe.

Eventually, the domestication of animals enabled their vastly greater power to be added to that of men. As the concept of animal power spread through various regions, horses, oxen, water buffalo, and elephants were harnessed for work in the fields. Three other sources of energy also became widely used: fire, wind, and moving water. The taming of fires set by lightning enabled people to keep warm and cook food. The heat energy from fire was also put to manufacturing use in the kilns (ovens) of potters and metalsmiths. Wind power was used to drive ships across the seas so that commerce could begin. Windmills for grinding grain into flour were constructed. In addition, mills powered by great wheels set in motion by water tumbling over a mill dam were perfected.

Agricultural productivity was greatly increased. Excess food production could be stored for future times of need. Food was one of the first forms of money. It could be traded for goods from other regions. Since men with their animals could grow enough food

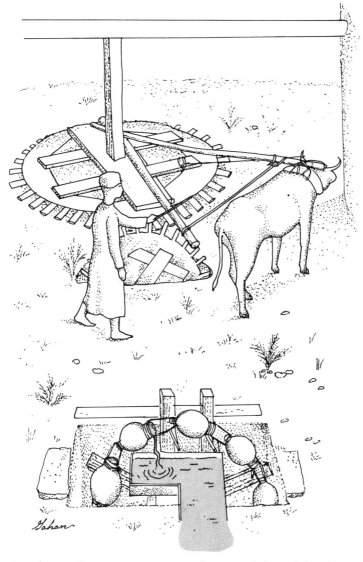

Animal power. Even in ancient times, people recognized the need for additional energy.

to feed many times their number, other people could work at occupations unrelated to agriculture. An artisan class developed, people who were skilled in various trades: potters, smiths, doctors, and others. A division of labor arose as a result of the increased availability of energy. Civilization grew more complex. Countries were made powerful by the energy at their disposal.

After several thousand years passed without the development of further sources of energy, the Scottish inventor James Watt perfected the steam engine and ushered in the age of industry. In countries that were once thickly forested, the demand for wood to fuel industrial fires grew so intense that the hills were stripped of trees. Today, in many parts of Italy, China, India, and the Middle East, the barren, eroded hills are a legacy of that unwise use. The onset of the industrial revolution in England and European countries greatly increased the demand for fuel energy to drive the newly developed machines. In these countries, coal, a fuel known and scorned for centuries, had to be exploited to satisfy the expanding energy needs of an industrializing society. For more than 200 years, well into the first half of the 20th century, coal powered the new machinery in the growing number of factories. Coal powered the trains and the steamboats. Coal heated the homes and cooking stoves. Coal was king of the fuels.

Petroleum products such as kerosene, gasoline, diesel oil, and fuel oil became increasingly popular during the 20th century. Before 1900, the greatest demand among petroleum products was for kerosene for lamps. Then the growth in automobile usage enormously increased the demand for gasoline. In addition, more and more homes gradually converted from coal furnaces to oil furnaces. By 1948 petroleum products along with natural gas accounted for more than 50 percent of the energy sources used in the United States. Oil had replaced coal as king of the fuels, and by 1973 oil and gas were the source of more than 75 percent of the energy needs of the United States.

What World Problems Can the Wise Use of Energy Solve Today?

Readily available energy is the key to maintaining our standard of living. Energy is also the key to solving the problems that trouble many peoples today. Overpopulation is a fact in countries such as India and China. Chronic food shortages have had their harmful effect on thousands of poor persons who daily struggle against starvation and famine. Providing the food to feed these hungry masses requires greater agricultural production than has ever been achieved.

Part of the success that has been achieved in the struggle against this problem lies in the "green revolution," the development of extremely productive varieties of grain. But these new forms of wheat and rice require large amounts of fertilizer to achieve maximum growth. Enormous amounts of energy are needed to manufacture the millions of tons of fertilizer necessary to keep the "green revolution" going successfully. Thus, energy is the answer to the problem of famine while the more basic problem of overpopulation is being brought under control.

Energy is the key to the conservation of our natural resources. In the past, production processes moved along a one-way street. Raw materials were shaped into finished products that were used and then discarded. Garbage dumps and refuse heaps have grown so huge that, in many areas, there will soon be no space left for disposal. One solution that has been proposed is to reprocess the waste into raw materials from which new products can be manufactured. This solution will require considerable expenditure of energy. The benefit of the plan is the conversion of a one-way street into a continuous cycle that will give us renewable resources.

Energy is the answer to relief from drudgery. Throughout much of the world, people still do most of their labor by hand. Many tedious, repetitive tasks can be performed by machines, provided that sources of energy are available to power them. Relief from such time-consuming effort would enable people to engage in more rewarding activities related to their real interests.

In the following chapters, we shall discuss aspects of the interwoven problems of energy, fuels, and the environment as they affect science, technology, economics, and society. These problems are your problems, and, if education is preparation for life, you need to examine and participate in solving them. It is only by facing reality that we can alter it and make our lives more satisfying.

QUESTIONS TO THINK ABOUT

1. What was the earliest type of power used by men?
2. What effect did the domestication of powerful animals have upon food production?
3. Name three other sources of energy.
4. What effect did the harnessing of more energy have upon society?
5. What was the first important industrial fuel?
6. Why was this fuel replaced by coal?
7. Why did petroleum replace coal as king of fuels?
8. Name three major world problems that the wise use of energy can help solve.
9. How would recycling conserve our national resources?
10. If our labor is handled by machines, how are our lives affected?

II

The Nature of Energy

The meaning of the word "energy" is difficult to understand because no one has ever seen energy. Energy must be thought of in terms of its effect on some form of matter, such as a speeding automobile or a rocket poised for imminent launching. Even the energy of light, which everyone assumes they can see, is invisible without some tiny specks of dust to show its presence. That is why the satellite photographs of interstellar space show it to be black. Although we know that light rays are streaming through the vast emptiness of space, the rays are unseen by us until they are reflected by interstellar dust. Similarly, radio and television waves must first react with the electrons in the receiver's antenna before they can be detected. Think of any form of energy. You can see or sense it only indirectly, after it has affected matter in some way.

Energy is usually defined as the ability to do *work*. Sometimes it is easier to think of energy as that which sets forms of matter into motion. Both definitions actually mean the same thing, because work is the result of matter in motion. Can you imagine any examples of work that do not include movement? Some students may argue that thinking, or playing a transistor radio, are examples of work without motion. Scientists have shown, however, that neither "thought" nor "radio sound" can be produced unless matter is in motion.

What Are the Many Faces of Energy?

Energy exists in various forms. Mechanical energy is the energy contained in moving bodies and is the most easily observed. All around us are machines and motors doing work in factories, homes, and farms. Automobiles, trucks, trains, ships, and air-planes transport people or freight.

Heat is also a form of energy, because it has the capacity to do work by putting matter in motion. A simple example is the rattling of a lid on a container of boiling water. Legend has it that young James Watt made practical application of this by inventing the steam engine in the 18th century. Actually, a Greek named Hero developed a small steam-driven turbine in the 1st century A.D. By the year 1698, Thomas Savery was granted a patent for a primitive steam engine for use in the English coal mines. Watt's real accomplishment was to improve the design of the early steam engines and make them truly practical. Today, the energy of heat is also used in internal combustion engines fueled by gasoline, diesel oil, or kerosene.

Radiant energy, the most obvious example of which is light, moves through space in the form of electromagnetic waves. These waves travel at the incredible speed of light, 186,272 miles per second. You may have used a prism to separate a beam of white light into a rainbow or spectrum of colors ranging from red to violet. Actually, this array of colors is but the small visible portion of the total electromagnetic spectrum of radiant energy. Physicists have shown that the essential difference between the colors is the length of their electromagnetic waves. The wavelength of red is the longest and that of violet is the shortest in the visible spectrum.

Later, it was learned that electromagnetic waves could be longer than the red, or shorter than the violet, but in all cases these new waves were invisible. The longer rays were called infrared, and could be felt as heat. Still longer waves were discovered and were called radio waves. Today, these are divided into longwave-

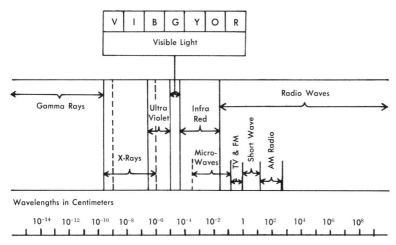

The electromagnetic spectrum. On this scale are listed the different forms of radiant energy. Their only difference is in the length of their electromagnetic wave. Overlapping occurs because sources of radiant energy produce a wide range of wavelengths. A gamma ray at point A in the diagram would be indistinguishable from and identical to an X ray at the same point.

AM, shortwave, television-FM, and microwave subsections. The waves too short to be seen include the ultraviolet rays, X rays, and gamma rays.

Radiant energy seems so weak that it appears incapable of doing work. However, you may have seen light start a radiometer spinning, its set of horizontal vanes, black on one side, silver on the other, whirling inside its glass enclosure. Light energy enables green plants to carry out photosynthesis, the chemical reaction by which carbon dioxide and water are converted into food and oxygen. It seems almost needless to say that radio waves are used to do the important work of communication. The X rays, gamma rays, and ultraviolet rays are used to do important work in medicine and scientific research.

Sound energy also travels in the form of waves, but its transmission requires a physical medium such as air, water, or solids. Sounds cannot traverse empty space. The awesome explosion of a star as it turns into a supernova occurs in an eerie silence. But on our planet, we use sound for communicating with each other by means of speech. Musical sounds entertain us. Loud sounds, such as the sonic booms caused by airplanes flying faster than sound, can break windows or do other damage. A useful form of sound energy, supersonics, has been put to work cleaning teeth in your dentist's office.

Electricity is our most useful form of energy because of the

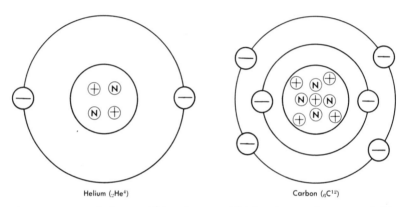

Helium ($_2$He^4)   Carbon ($_6$C^{12})

Atoms are composed of protons ($\oplus$) and neutrons ($\otimes$) in a dense nucleus, surrounded by electrons ($\ominus$) moving rapidly in their orbitals (not drawn to scale).

convenience with which it can be regulated, transmitted, and transformed into other forms. Plug your favorite electrical appliance into the wall socket, turn on the switch, and you have light, sound, heat, or mechanical energy at your fingertips.

Electrical energy has its origin in the forces exerted by electrically charged particles found in the atom: electrons and protons. You can demonstrate this, and prove to yourself some of the laws of electricity, by doing simple static-electricity experiments. On a cool dry day, rub a plastic or rubber comb on a piece of wool. The comb will be able to attract small, light objects such as bits of paper, feathers, or thread. Use a second comb to experiment with objects already attracted to the first comb. You will notice forces of repulsion as well as attraction.

The presence of two kinds of force indicates that there are two kinds of electrical charge: positive and negative. Careful observation indicates that like charges repel, and unlike charges attract. This is the law of static electric charges. It is necessary to use insulating materials that do not conduct electricity to develop

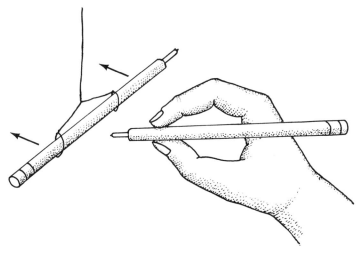

Static electricity. Suspend one plastic ballpoint pen with thread as shown. Bring a second pen up to it after charging both pens with a wool cloth. You will observe a force of repulsion.

such static electric charges. Metallic materials tend to be electrical conductors because they permit electrons to flow easily over their surface. Thus, electric currents are the result of a flow of electrons, the negatively charged subatomic particles. Similarly, negative static charge is the result of a large number of electrons gathered on the surface of an insulated object. Rubbing electrons off a surface produces a positive charge because there is now an excess of protons.

Chemical energy is the energy stored in the electron arrangements of the molecules of such substances as fossil fuels, wood, explosives, and food. The energy can be released when the molecules react chemically. During the reaction, the molecules, which are clusters of atoms bonded together by the electrons shared between them, rearrange their atomic composition into a more stable structure that requires less energy to hold it together. It is the extra energy that is released as chemical energy.

Since electrons are also responsible for producing electrical energy, it is not surprising that one application of chemical energy is the battery. The dry cell is an arrangement of chemicals that react to release electrons at a constant rate. This results in a steady current of electricity until the chemicals are used up. In rechargeable storage batteries, an electric current can be run back into the battery, reversing the usual reactions and causing the original chemicals to be re-created in the battery, thus restoring its useful life.

Atomic energy or nuclear energy stems from the force that holds atomic nuclei together. If this force were not one of the strongest known in nature, the atomic nucleus would explode, since packed together in it are protons, all with positive charges of electricity. These particles should repel each other with such force that every atomic nucleus ought to burst. But they do not, and physicists believe the reason is the binding energy of a mysterious nuclear force.

When an atomic nucleus splits, the energy holding it together is released in the form of heat and gamma rays. The important thing is that the amount of energy released in nuclear reactions is thousands of times greater than that given off in chemical reactions. Thus, nuclear energy is the world's best hope for an eventual solution to our energy needs.

Physicists often speak of energy in terms of its kinetic or potential nature. Energy is kinetic when it is associated with moving bodies. The motion may have a purpose, such as a machine doing work, or it may be of a more random nature, such as a gentle wind or flowing stream. In any case, whenever a physical body is in motion, be it as large as a planet or as small as an electron, kinetic energy is at work.

Potential energy represents the hidden face of energy. It describes those situations in which we know that a physical object may be capable of work or motion even though it is temporarily at rest. An unfired bullet still has the capacity to move because of the chemical energy in the gunpowder. Since the bullet is not actually moving, the energy is potential in nature. A car parked on a steep hill also has potential energy because of its position. In San Francisco, a city famous for its hills, motorists are warned to park with their wheels turned in toward the curb as a safety precaution. In case of brake failure, the car will jam up against the curb and sidewalk instead of rolling wildly down the slope.

Elastic materials, such as steel springs, rubber bands, or wooden diving boards, can possess potential energy by the distortion of their forms. Small mechanical toys are often operated by means of the potential energy stored in a wound-up spring.

Less obvious examples of potential energy are the energy in chemical bonds and the energy locked in atomic nuclei. Fuels, for example, are materials that possess abundant potential energy. Fossil fuels contain chemical energy. Nuclear fuels are sources of atomic energy. In all these instances, the energy is potential in nature because it is latent, stored, or unused.

What Transformations Can Energy Undergo?

Energy is constantly being changed from one form to another. This is a well-established fact, easily proven by everyday experience. The transformation of energy is most readily accomplished using electricity. A light bulb converts electrical energy into light.

Toasters change electricity into heat. Motors change electricity into mechanical energy. Earphones and loudspeakers convert it into sound.

Physicists have determined that any form of energy can be transformed into any other form of energy by the appropriate energy-conversion device. For example, coal is burned at a power plant, changing its chemical energy into heat energy. The heat energy is used to make steam, which in turn is converted into mechanical energy by rotating a turbine connected to an electrical generator. The generator changes the mechanical energy into electricity, which travels along wires to the users. There, the electrical appliances act as energy-converting devices as previously described.

What Is Meant by the Conservation of Energy?

Careful analysis of the transformation of energy shows that energy is neither created nor destroyed during the process. This finding has led to a fundamental law of science, the conservation of energy, which states that the amount of energy entering an energy-converting device is equal to the amount of energy leaving it. Another way of stating the law is that the total amount of energy in an energy-converting system is fixed or constant.

Although energy cannot be destroyed, it can be wasted. This happens whenever energy is not used for the desired purpose. Engineers attempt to build energy-conversion devices that waste the minimum amount of energy. The ability of the device to use energy in a purposeful way is an indication of its efficiency. For example, an incandescent lamp wastes electricity by changing much of it into heat instead of light. A fluorescent lamp is more efficient because it converts most of the electricity into light and wastes only a little in the form of heat.

One result of the inevitable energy loss through wastage is the impossibility of building a perpetual-motion machine. Once set in motion such a device would continue to run indefinitely with no further energy input. Such a machine would have to be 100 percent efficient. The inevitable heat loss prevents this from ever being achieved. This means that all the "brilliant" ideas for accomplishing perpetual motion are doomed to failure. In one arrangement, a storage battery is connected to an electric motor, which in turn rotates a generator producing an electric current. The generator is connected to the storage battery to keep it recharged. Unfortunately, the energy available at the end of each cycle is always less than that put into the beginning of the cycle because of heat loss. The machine must run down eventually.

The principle of energy conservation is the first law of thermodynamics. This law describes the relationship between work and heat by stating that they are equal to each other when measured in the same units. In other words, a certain amount of work can be transformed into an equal amount of heat, or conversely, a certain amount of heat can be transformed into an equal amount of work.

The inevitable loss of energy that occurs in such energy transformations is described by the second law of thermodynamics, which states that heat always flows from a hotter region to a colder region. The result of this is that when the temperature has become uniform, heat stops flowing. The system has become stable, which means fixed, steady, or unchanging. Physicists have long observed the tendency of all forms of matter and energy to become as stable as possible spontaneously.

The stability of any form of matter depends on the amount of potential energy it contains. The term often used to describe this is energy level. The higher the energy level of the material, the more unstable it is and the more likely it is to undergo some type of change. For example, a truckload of nitroglycerine contains a much higher energy level than a truckload of sand. Therefore, the driver of the truckload of nitroglycerine receives a much higher wage to compensate for the danger inherent in transporting so unstable a material. If given the opportunity, the nitroglycerine will become more stable by exploding, thereby losing most of its potential energy and enabling it to reach a much lower energy level.

This tendency of substances to become more stable is the driving force for all the changes, chemical, nuclear, and physical, that occur in nature. Matter will always lower its energy level spontaneously but can never increase it unaided. This is the meaning of the second law of thermodynamics: heat flows only to cooler regions, never the reverse. An electric iron begins cooling off immediately after it is unplugged because it loses heat to the surrounding air; but no one has ever observed the heat in the air spontaneously flowing into the iron, raising its energy level to the pressing temperature for a wool suit. This will not happen without aid, which we give by simply replugging the iron into the electric outlet, an external source of high potential energy. The simple significance of the thermodynamic laws is that, in ordinary transformations, energy is never destroyed, but always wasted.

How Do We Measure the Quantity of Energy in Work and Power?

It is often necessary to measure quantities of energy accurately. This is important when buying fuels in large amounts to produce electric power or do other kinds of work. The engineers in charge of an electric generating plant can tell from experience how much power they will need to produce at a given time. By their calculations, they can then determine how much fuel they must use in order to accomplish this.

Work, the result of setting matter in motion, is done by the application of kinetic energy. This requires that a force be applied to an object to move the object a certain distance. The work (W) is the product of the force (F) multiplied by the distance (D), or $W = F \times D$. In the metric system, the unit of force is the dyne or newton and the unit of work is the erg or joule. In the English system, the unit of force is the pound and the unit of work is the foot-pound.

SAMPLE PROBLEM

How much work must be done to lift a 50-pound crate to a height of 10 feet?

SOLUTION

Substituting in the formula $W = F \times D$, we obtain $W = 50$ pounds $\times 10$ feet. Solving, $W = 500$ foot-pounds.

Power is the measure of how fast work is done. A physicist would say power is the rate of doing work. The formula for calculating power (P) is equal to the work (W) divided by the time (t), or $P = W/t$. In the metric system, the unit of power is the watt, equal to one joule per second. The horsepower, equal to 550 foot-pounds per second, is the most common unit of power in the English system. The formula for calculating horsepower (HP) is equal to the work divided by the product of 550 multiplied by the time (t), or $HP = W/550 \times (t)$.

SAMPLE PROBLEM

How many horsepower are developed when a force of 100 pounds is used to lift a heavy crate 55 feet in 5 seconds?

SOLUTION

Substituting in the formula $HP = W/550 \times (t)$, first calculate the work (W) done by substituting in the formula $W = F \times D$; we obtain $W = 100$ pounds $\times 55$ feet. Solving, $W = 5500$ foot-pounds. Having found W, substitute its value in the formula $HP = W/550 \times (t)$, obtaining 5500 foot-pounds/550 $\times$ 5 seconds. Solving, $HP = 2$ (horsepower).

The watt is also the unit of electrical power and is equal to one joule of electrical work per second. In electrical units, watts (W) are the product of volts (V) multiplied by amperes (I), or $W = V \times I$. The volt is a measure of the work per unit of charge. The ampere measures the rate at which electrical charge moves through the circuit, or how many charges per second are flowing past a point in the circuit. Therefore, Watts = work/charge $\times$ charge/second, or—simplifying the formula because the charges cancel each other—Watts = work/second, which means watts measure the rate of doing work. A kilowatt is equal to 1,000 watts. A kilowatt-hour (Kwh) is the use of one kilowatt of power for a time duration of one hour. As a formula, Kwh = watts $\times$ hours/1,000.

SAMPLE PROBLEM

How many watts of power are used by an electric bulb that requires 2.5 amperes of current on a 120-volt line?

SOLUTION

Substituting in the formula $W = V \times I$, we obtain W = 2.5 amperes $\times$ 120 volts. Solving, W = 300 (watts).

SAMPLE PROBLEM

How many kilowatt-hours are used by a 100-watt bulb left on for 20 hours?

SOLUTION

Substituting in the formula Kwh = watts $\times$ hours/1,000, we obtain 100 $\times$ 20/1,000. Solving, Kwh = 2 (kilowatt-hours).

SAMPLE PROBLEM

a. How many watts of power are used by a 7.5-ampere air conditioner on a 120-volt circuit?
b. How many kilowatt-hours does it use if it runs for 10 hours?

SOLUTION

a. Substituting in the formula $W = V \times I$, we obtain W = 7.5 amperes $\times$ 120 volts. Solving, W = 900 (watts).

b. Substituting in the formula Kwh = watts × hours/1,000, we obtain Kwh = 900 × 10/1,000. Solving, Kwh = 9 (kilowatt-hours).

SAMPLE PROBLEM

If electricity costs $0.05 per Kwh, how much did it cost to run the air conditioner?

SOLUTION

The cost of operating an electrical appliance is the product of the number of kilowatt-hours multiplied by the cost of a kilowatt-hour. Therefore, we obtain that the cost equals 9 Kwh × $0.05/Kwh. Solving, cost = $0.45.

How Do We Measure Heat Energy?

Energy is the ability to do work, and as more energy is available, more work can be done. Physicists have developed methods and formulas for measuring amounts of the different forms of energy in terms of the work that energy can do. Let us see how this idea is applied to measuring an amount of heat energy. The effect of heat is to make objects warmer by speeding up their molecules. You can begin measuring the work done by using a thermometer to note the increase in temperature. In addition, you must take into account the size of the object that is being heated, since it will require more heat to warm a large body than a small one.

In scientific investigations, the physicist uses a device known as a calorimeter to measure heat energy. The calorimeter consists of a thick-walled insulated container filled with a known weight of water. A sealed combustion chamber containing oxygen enables test samples of different materials to be burned. The heat from the burned sample warms the water, and a sensitive thermometer records the increase in temperature.

One unit of heat measurement is the calorie. A calorie is the amount of heat needed to raise the temperature of one gram of water one degree Celsius (formerly centigrade). Since this amount of heat is very small, the kilocalorie or large Calorie, is commonly used. The Calorie stands for 1,000 times more heat than the calorie, because it is the amount of heat needed to raise the temperature of 1,000 grams of water one degree centigrade. Dieticians and nutritionists use the Calorie to measure the energy content of foods.

Although most countries use the metric system and Calorie, some nations still use the English system and measure heat in British thermal units (Btu's). A Btu is the amount of heat needed to raise the temperature of one pound of water one degree Fahrenheit.

TABLE OF UNITS

System	Force	Work	Power	Heat
Metric centimeter gram second	dyne	erg (dyne-cm.)	erg/second	calorie (small)
Metric meter kilogram second	newton	joule (newton-m)	watt (joule/second)	Calorie (large)
English foot pound second	pound	foot-pound	foot-pound/second 1 horsepower = 550 foot-lbs/second	Btu (British Thermal) Unit

In each case, the formula for calculating the quantity of heat that water gains or loses is:

$$\Delta H = w \times (\Delta T) \text{ where}$$

ΔH is the change in heat content (energy)
w is the amount of water by weight
ΔT is the change in temperature. $T = T_1 - T_0$ where

T_1 = final temperature
T_0 = original temperature

In the metric system, the formulas would read:

a. H (calories) = grams of water × change in degrees C.
b. H (Calories) = kilograms of water × change in degrees C.

In the English system, the formula would read:

H (Btu's) = pounds of water × change in degrees F.

SAMPLE PROBLEM

What quantity of heat is needed to raise 100 pounds of water from 50 degrees F. to 60 degrees F.?

SOLUTION

Substituting in the formula Btu's = pounds of water × change in degrees, we obtain Btu's = 100 pounds × (60 − 50). Solving, Btu's = 1,000.

SAMPLE PROBLEM

a. What quantity of heat is needed to raise 500 grams of water from 10 degrees C. to 30 degrees C.?
b. What quantity of heat is needed to raise 3 kilograms of water from 15 degrees C. to 20 degrees C.?

SOLUTION

a. Substituting in the formula calories = grams of water × change in degrees C., we obtain calories = 500 × (30 − 10). Solving, calories = 10,000.
b. Substituting in the formula Calories = kilograms of water × change in degrees C., we obtain Calories = 3 × (20 − 15). Solving, Calories = 15.

SAMPLE PROBLEM

What will be the final temperature when 1,000 pounds of water at 50 degrees F. are heated by 5,000 Btu's?

SOLUTION

Transpose the formula Btu's = pounds of water X change in degrees F. to change in degrees F. = Btu's/pounds of water. Substituting in the formula, we obtain change in degrees F. = 5,000 Btu's/1,000 pounds. Solving, change in degrees F. = 5. Since change in degrees F. = $T_1 - T_0$, we must transpose this formula T_1 = change in degrees F. + T_0 to find the final temperature. Substituting in the formula we obtain $T_1 = 5 + 50$. Solving, $T_1 = 55$ degrees F.

SAMPLE PROBLEM

a. A 1-ounce sample of coal raises the temperature of 175 pounds of water 5 degrees F. when burned. How many Btu's are produced when a pound of the same coal is burned?
b. When a ton of the same coal is burned?

SOLUTION

a. First calculate the Btu's in the ounce of coal. Since there are 16 ounces in 1 pound, multiply that answer by 16 to calculate the Btu's in 1 pound of coal. Substituting in the formula Btu's = pounds of water X change in degrees F., we obtain Btu's = 175 X 5. Solving, Btu's = 875 in 1 ounce of coal. Multiplying by 16 to obtain the Btu's in 1 pound of coal, we obtain Btu's = 875 X 16. Solving, Btu's — 1 pound of coal = 14,000.
b. There are 2,000 pounds in 1 ton. Multiply the Btu's in 1 pound of coal by 2,000 to obtain the Btu's in 1 ton of coal. Btu's in 1 ton of coal = 14,000 X 2,000. Solving, Btu's = 28,000,000.

With practice, it becomes easy to calculate how much energy it takes to heat water or do other kinds of work. It is much more difficult to obtain that energy in the first place. Thus, it becomes very important to know the sources from which the necessary amounts of energy can be found. In the next chapter, we will discuss some of the important sources of energy that can be utilized in solving the power crisis facing all of us.

QUESTIONS TO THINK ABOUT

1. How does energy show its presence?
2. What is the usual definition of energy?
3. Name seven forms of energy.
4. Give two examples of work done by mechanical energy.
5. What has been the most important invention using heat energy?
6. What form of energy is light?
7. What electromagnetic rays are invisible?
8. Give some examples of work done by radiant energy.
9. What is required for the transmission of sound energy?
10. What makes electricity our most useful form of energy?
11. What is the law of static electricity?
12. Devise an experiment that uses static electricity to perform work.
13. What is an electric current?
14. What is meant by chemical energy?
15. How does a battery work?
16. What happens when an atomic nucleus splits? What is this energy called?
17. Give an example of a substance that has kinetic energy.
18. Define potential energy. Give two examples.
19. What is the principle of the transformation of energy?
20. What is required in order for energy to be transformed?
21. Cite two examples of energy transformation.
22. What is the law of the conservation of energy?
23. When heating water on a stove, how is some energy wasted?
24. What is meant by the "efficiency" of a machine?
25. Why won't perpetual motion machines work?
26. State the second law of thermodynamics.
27. What is the principle behind all physical and chemical changes?
28. How much work must be done to lift a 40-pound valise to a height of 3 feet?
29. How much work is done when a force of 30 newtons is used to lift a 15-kilogram box to a height of 2 meters?
30. How many horsepower are developed when a force of 300 pounds is used to lift an automobile 11 feet in 6 seconds?
31. How many watts of power are used by an air conditioner that requires 12.5 amperes of current on a 120-volt line?

32. How much would it cost to operate the above air conditioner for a period of 100 hours if 1 kilowatt-hour costs $0.05?
33. What quantity of heat is needed to raise 500 pounds of water from 70 degrees F. to 85 degrees F.?
34. What quantity of heat is needed to raise 200 grams of water from 10 degrees C. to 20 degrees C.?
35. How much heat energy is released when a 100-gram peanut-butter sandwich burns in a calorimeter and heats 70 kilograms of water from 20 degrees C. to 25 degrees C.?

III

Sources of Energy

In this chapter, we shall discuss the many ways of producing and using energy. Some sources of energy have been with us for thousands of years, others are less than one hundred years old, and still others are dreams yet to be fulfilled. In all cases, however, effective use of these energy sources involves harnessing the natural forces that are available to us. Because of their prominent place in our lives, the fossil fuels (coal, petroleum, and natural gas) will be discussed in later chapters. The discussion of nuclear fuels in this chapter will be expanded upon later in the book.

How Can Solar Energy Be Used?

Energy comes to us from many sources; however, man has always recognized that the sun and the reactions constantly occurring on the sun are our primary source of energy. Many ancient civilizations, such as Egypt at the time of the pharaohs, thought the sun was a god, and it's easy to see why. During seasons when the sun shone high in the sky and the land was warm, crops would flourish and provide food. When the sun was low, however, and daylight hours were few, light and radiant energy were insufficient, and the land remained barren. In ancient times, people of many cultures were afraid of darkness. They did not understand how the motions of the earth create night and day and the seasons. Even today, children often fear darkness until they realize why it occurs.

We know that the earth rotates (spins) on its axis, the imaginary line running from the North Pole to the South Pole. At any one time, half the earth faces the sun and is in daylight while the other half is experiencing night. As you probably know, the earth makes one complete rotation every twenty-four hours.

As rotation is taking place, another type of motion is occurring. The earth is swinging around the sun with an almost circular motion, known as its revolution. The earth revolves around the sun every 365¼ days.

Seasons occur on earth because its axis is tilted. When the earth is on one side of the sun, the North Pole is tilted toward it. This gives the Northern Hemisphere, where we live, more direct sunlight and more hours of daylight. The season is then summer. At the same time, it is winter in the Southern Hemisphere, where there are fewer hours of sunlight each day and the sun is lower in the sky.

Six months later, the earth is on the other side of the sun. Now the South Pole is tilted toward the sun while the North Pole is tilted away from it. The Southern Hemisphere now has more direct sunlight and it is in summer, while the Northern Hemisphere is in winter.

Despite the fact that weather conditions change every day, and that our planet is about 93,000,000 miles away from the sun, the amount of sunlight reaching the earth is enormous. The surface of the sun reaches a temperature of 6,000° C. Radiant energy (sunlight) reaches the earth at an average rate of two calories per minute for every square centimeter of surface area. This amount of energy is equivalent to burning over 1,000,000,000,000,000 pounds of coal each day. At that rate, it would take less than an hour to supply the yearly power needs of the United States if all this energy could be used.

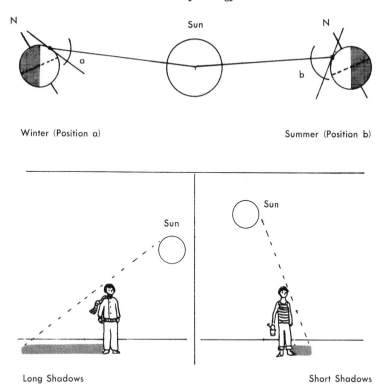

Revolution. *When the earth is in position A, the sun is lower on the horizon and its rays strike less directly at Point 1. The North Pole is in winter and the South Pole is having summer. When the earth is at position B, the sun is more directly overhead, and its rays strike with greater strength at Point 1. Now the North Pole is in summer and the South Pole is having winter.*

The power of the sun can be easily demonstrated with a magnifying glass. The lens is so designed as to concentrate the sun's rays that pass through it at a point some distance away from it. If you hold the magnifying glass steady with the lens facing the sun and place a piece of paper such as a tissue at the point where the rays are focused (known as the focal point), you will see the paper soon catch fire.

Scientists today are looking toward capturing solar energy and substituting its use for that of fossil fuels such as coal, petroleum products, and natural gas. Solar energy has already been used to cook food, heat homes, and purify water by distillation. It is hoped that such energy can be used on a large scale to irrigate land and to power telephone lines as well.

To create the high temperature necessary to operate solar boilers and furnaces, curved (parabolic) mirrors are arranged to collect the sun's rays and focus them onto a black surface. A black surface is used because the rays are absorbed by it, producing heat. If the surface were white, the light energy would be reflected from the surface and less heat would be produced. The same principle is applied when people in cold climates wear dark clothing to keep warm and people in warm climates wear light-colored clothing to stay cool.

In the world's largest solar furnace, at Odeille in the French Pyrenees, temperatures as high as 6,000° F. may be reached quickly. Many flat mirrors called heliostats are turned at such an angle that the sun's rays are reflected from them toward the huge parabolic mirror. The heat produced has been useful in making ceramic materials and testing metals. Similarly, a solar furnace in San Diego, California, reaches a temperature of 8,500° F. A great advantage of such furnaces is that the articles being tested or made can be kept in a much purer condition than is possible with conventional furnaces.

Houses and water may be heated with this energy source. A flat plate collector is mounted on a roof (the part that slopes south receives the most radiation). Blackened tubes in the collector carry water or air, which is heated by the sun's rays. As a result of the so-called greenhouse effect, the glass above the tubes lets the sun's rays pass and heat the tubes, but does not allow the heat waves coming from the tubes to escape to the atmosphere. The heated water may be stored in insulated containers. The hot air may be used to heat insulated crushed rock. When heat is required, cold air is blown through the heated rock to warm it before it enters the home. This method is being used in the southern and southwestern parts of the United States.

In a similar manner, electricity may be produced by solar power. The radiant energy can heat water to boiling, and the

Solar panels power a radio repeater station near Bozeman, Montana.

steam may turn the turbines necessary to generate an electric current. The turbines may also help in the irrigation of farmlands by serving as pumps for water. Such a project was first attempted in 1912 on the Nile River in Africa. It has also been tried in Arizona and California, but without much economic success.

Water may be purified by solar energy. The water is placed on a pan that is blackened to absorb heat. The evaporated water condenses on a glass roof placed above the pan, and the water runs off the roof to a collection point. Purification of seawater has been accomplished this way in many parts of the world.

In 1954 the Bell Telephone Laboratories developed the first solar batteries, using crystals made of silicon with traces of other elements; electricity is generated when light strikes them. An advantage of solar batteries is that the crystals never wear out. They have been applied in flashlights and in powering telephone lines. They have also been part of the power equipment on space vehicles such as satellites. In the future, satellites may be designed to capture the sun's energy and transmit it back to earth via electromagnetic waves generated in the satellite.

Despite the many plans to make solar energy available to man, its large-scale use appears to be years away. Several obstacles must first be overcome. Although much radiant energy strikes the earth, it is thinly spread out. Power plants would have to be very large in order to capture enough sunlight. It is obvious, too, that the amount of radiation varies from day to day. No effective method of storing large amounts of solar power has been developed; thus it would be necessary to install additional systems for energy production during long periods of cloudy weather. Space satellites can overcome problems of weather, but the cost of such satellites and the solar batteries within them is very great. Scientists and engineers, however, believe that someday these difficulties will be overcome.

How May Wind Power Be Harnessed?

Men have been harnessing the power of the wind for many hundreds of years. It is known that windmills were used in Persia (modern-day Iran) in 644 A.D. Many large windmills still stand in the Netherlands and other northern European countries. Many

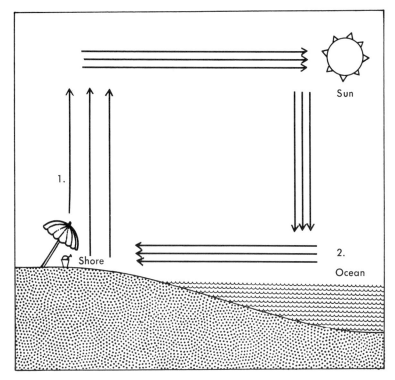

The origin of a breeze. On a hot day at the beach, you will probably feel a breeze coming from the water. The sun heats the land faster than it heats the water. The heated air rises (see 1). The rising air creates a partial vacuum, which leads to the breeze of cooler air rushing ashore to replace it (see 2).

can also be seen in the American countryside. Wind power has been used as a source of electric power in Denmark since 1890.

Generally, winds originate because of the sun's unequal heating of the atmosphere at different points on the earth's surface. For example, near a shoreline, the air over the water will not heat nor cool as quickly as air over the land. The difference in the air temperature creates a wind current that can be very strong.

Other important factors that influence the strength and direction of the wind are altitude (mountains and valleys) and latitude (distance north and south of the equator). Over the years, several designs for windmills have been made. In horizontal mills, the sails are mounted on a vertical axis. In vertical mills, the axis is horizontal. The mill can be turned, if necessary, so that the sails face the wind. Through a system of gears, the energy of the wind is converted to mechanical energy within the mill. The first windmills were used to grind grain. In addition to producing electricity, windmills are now used to pump water from wet land, to saw timber, and to make paper. Some scientists have proposed that windmills be floated onto a large lake or an ocean, where they would not be an eyesore, yet would produce large amounts of electricity.

How Can Geothermal Energy Be Used?

Scientists today believe our planet was formed from hot gases that separated from the sun billions of years ago. Because of the radiation of its heat into outer space, the earth eventually cooled, first liquefying and then partly solidifying to form an outer crust. Beneath the crust, however, the earth is still hot liquid, the heat partly originating from radioactive elements within it. Some energy experts believe that this heat, known as geothermal energy, can serve as the primary source of energy in the future.

Heat from the earth is being used today in some countries. In Lardarello, Italy, is a natural steam field where underground water is turned to steam by the earth's heat. The steam has been powering turbines to generate electricity since 1913. Similarly, geysers (hot springs that periodically throw off water) are driving generators in California and are expected to be harnessed further in the future.

In some parts of the world, the hot rocks lie only 1 to 5 miles down. One ambitious plan has been put forth to drill pairs of holes deep into the hot portion of the crust. Cold water would be sent down one hole, where it would be heated to steam and rise up the other hole. The steam would drive turbines and then be recycled back to the first hole. The advantages of utilizing geothermal energy are its low cost and its lack of pollution.

How Is Waterpower Used?

The energy of moving water has been applied for thousands of years. Ancient Egyptians placed a waterwheel on the Nile River to irrigate their fields. The current turned the wheel and caused buckets of water to be raised and poured into a trough that led to the fields. Just as old is the method of adapting waterpower for processing grain. In the 1600s the Pilgrims in America used it for grinding corn.

With the development of turbines in the 19th century, water began to be used to generate electricity. The first hydroelectric station was built in Appleton, Wisconsin, in 1882. In the United States today, about 30 percent of the potential waterpower has been developed, based on the construction of dams.

Dams employ the force of falling water to produce electric power. In addition, the dams may also help in irrigation and flood control, improve navigation, and maintain and develop recreation lands and forests. In 1933 an ambitious program of dam building was begun when the Tennessee Valley Authority (TVA) was established by the United States Congress. The TVA has built more than twenty dams along the Tennessee River and its tributaries. It sells electric power to both local electric companies and the U.S. government. Dams have been built on many of the major rivers in the United States.

Among the disadvantages of hydroelectric power is the fact that seasonal variations constantly change the flow characteristics of rivers, making it difficult to extract the maximum benefit from the

A hydroelectric dam.

dam. Also, it is difficult to store the energy that can be produced in peak periods; however, a reservoir type of dam can store water for release as required.

In addition to dams, tides and ocean currents can be tapped to generate electricity. In the Bay of Biscay, France, tidal currents are being used. Canada and the United States are developing plans to capture the tidal power in the Bay of Fundy, near the eastern border of the two countries. Turbines with reversible propellers can be operated by both incoming and outgoing tides. In Great Britain it has been proposed that floating generators be placed in the North Atlantic Ocean to harness the power of the waves. It is believed that this source alone might satisfy that country's present power requirements.

What Is Atomic Energy?

Atomic energy refers to the energy that is released when the nuclei of atoms combine or separate to form new elements. Atomic energy differs from ordinary chemical energy, which originates in the electronic forces that form molecules from individual atoms. The forces that hold the particles of the nucleus together are much stronger.

The French scientists Henri Becquerel, Marie Curie, and her husband Pierre Curie were among the first to study the energy given off by naturally radioactive elements such as radium. In the late 19th century, they investigated powerful rays given off by these substances. Three rays were discovered and named alpha, beta, and gamma after the first three letters of the Greek alphabet. In 1905 the American physicist Albert Einstein proposed a theory of relativity, which indicated that matter could be transformed into energy. Since the total atomic mass should decrease during a nuclear reaction, the Einstein equation gave the amount of energy which would result from that decrease.

As the years passed, ways were developed to create reactions that split (fission) atoms and to control the nuclear particles involved in them. In the late 1930s several scientists believed that such fission reactions had the potential to release vast amounts of energy; however, controlling the reactions would be extremely difficult. Fortunately, the great Italian physicist Enrico Fermi and other scientists developed the necessary techniques.

At the time, the world was embroiled in World War II. The Manhattan Project was set up to develop an atomic bomb before the enemy could do so. Many European scientists who fled the Nazi persecution of that time came to the United States to help that effort. On July 16, 1945, the first atomic bomb was exploded near Alamogordo, New Mexico. Its blast was equivalent to the explosion of 34,000,000 pounds of TNT. Less than a month later, on August 6, an atomic bomb demolished the city of Hiroshima,

Japan, and 66,000 people were killed. Three days later, 39,000 were killed when a similar bomb exploded in Nagasaki, Japan. World War II ended the next day when the Japanese forces surrendered.

These bombs were based upon the use of rare heavy elements, uranium and plutonium, as the explosives. After the war, a new type of bomb was developed that used a form of hydrogen, an abundant and light element, as its explosive. This hydrogen bomb, based upon the union (fusion) of hydrogen atoms, had far greater power than the previously developed fission bombs. The fusion reaction is similar to that which occurs on the sun, where 40,-000,000,000 tons of hydrogen react every minute to produce its heat energy.

Most scientists believe that controlled fusion may well be the world's major future energy source. Meanwhile, nuclear fission reactors that utilize uranium are set up in several countries to produce energy. It is estimated that the oceans contain enough hydrogen (water's formula is H_2O) to provide energy for an almost unlimited time. The major factor that is delaying development of these energy sources is the difficulty of controlling the reactions.

Control of nuclear reactions is necessary so that energy is released slowly and safely, rather than all at once. The United States plans to have more than one hundred nuclear reactors in operation in the not too distant future, supplying 30 percent of the nation's electrical power. The reactors would contain uranium and plutonium as fuels, the atoms of these metals slowly splitting to form lighter elements, releasing energy during the process. The energy may heat water to steam, which then runs turbines and generators to produce electricity.

In addition to use in generating electricity, nuclear fuels help propel submarines and other ships. In medicine, radioactive chemicals manufactured in the nuclear reactor have been applied

COURTESY ATOMIC INDUSTRIAL FORUM, INC.

The Fort Calhoun Nuclear Power Station, near Omaha, Nebraska, has a capacity of 455,000 kilowatts of electricity.

to diagnose diseases and destroy cancers. Other radioactive chemicals have been employed in studying the structure of both living things and industrial materials.

It will be more difficult to control reactions involving hydrogen atoms because the materials of production are harder to work with and the temperatures necessary to begin the process are much higher. However, research is going on in several countries, including the United States, to develop methods of utilizing this energy.

In 1955 the first International Conference on the Peaceful Uses of Atomic Energy met in Geneva, Switzerland. At the same time, the United States began its Atoms For Peace program, which gave nuclear materials and other forms of assistance to nations wanting to apply atomic energy for peaceful purposes. The Western European countries of France, West Germany, Italy, the Netherlands, Belgium, and Luxembourg formed Euratom, an organization devoted to building nuclear reactors for research. Through recent years, nations have been seeking ways to cooperate in the peaceful application of atomic energy.

QUESTIONS TO THINK ABOUT

1. What is our most important energy source?
2. Describe the earth's rotation. How long does it take for each full rotation?
3. How long does one revolution of the earth take?
4. Why is it cold in winter? Why is it warm in summer?
5. When it is summer here, why is it winter in Australia?
6. To what uses has solar energy already been put?
7. Describe how a solar furnace is made.
8. How can the "greenhouse effect" help keep homes warm?
9. What are solar batteries?
10. What obstacles block greater use of solar power?
11. Name two devices for harnessing wind power.
12. To what uses have windmills been put?
13. Define geothermal energy.
14. What are geysers, and how can they be used?
15. What are the advantages of harnessing geothermal energy?
16. How was waterpower first harnessed?
17. What is the most important way waterpower is used today?
18. What other benefits can be derived from the construction of a dam?
19. What is the advantage of a reservoir type of dam?
20. In addition to rivers and streams, name two major sources of potential waterpower.
21. How does atomic energy differ from ordinary chemical energy?
22. What do radioactive elements do?
23. What did Einstein's theory of relativity show?

24. What were the first important applications of atomic energy?
25. What fuel is used in atomic bombs and reactors?
26. What is the basis of the hydrogen bomb?
27. Why is nuclear fusion possibly our major future energy source?
28. What are some uses of nuclear energy?

IV

Introduction to Fossil Fuels

Fossil fuels are by far our most important current sources of energy. The power from coal, oil, and natural-gas products enables the machines in factories to run, planes to fly, cars and trucks to roll, and ships to sail. In addition, we use these fuels to heat our homes, schools, and office buildings and to cook our food.

In this section, we shall discuss how the fossil fuels formed, how they are processed, and the uses for which they are best suited.

What Were Fossil Fuels Long Ago?

Millions of years ago, the earth was much warmer than it is today. Even the North and South poles were free of ice and snow. The warm climate and abundant water released by the melted ice formed huge swamps and shallow inland seas. These conditions were highly favorable to the growth of living things. Everywhere, plants and animals multiplied in larger and larger numbers. Forests and jungles were thick with trees, giant ferns, and other plants. Animals of every size and form swarmed in the warm waters and along their shores. Each year, many of these plants and animals died, and their bodies formed the fossil fuels we use today.

As each plant or animal died, its body sank into the swamp or dropped to the bottom of the sea. Bacteria, the cause of decay, gathered on the dead remains and, using them for food, began to oxidize the fleshy substances. Before they could complete their job, the heavy rainfalls common in that moist climate eroded the huge mountains that existed then. The mud and sand washed down by these torrential storms swept over the swamps, and swift rivers carried the silt out into the sea. These sediments buried the partly decayed remains and, by cutting off the supply of oxygen, prevented the bacteria from finishing their task of decomposition. In this way, the materials of these dead plants and animals were preserved for future generations to use as fuel, instead of being broken down and returned to the earth as simple compounds.

In time, the clay and sand hardened into different kinds of sedimentary rock. The clay formed shale, and the sand grains were cemented by minerals into sandstone. As the ages passed, more sedimentary rocks formed on top of the older layers. Dense rocks such as limestone developed as the lime shells of millions of tiny sea creatures were deposited on the floors of these shallow seas. Volcanoes also aided the rock-building process. Their eruptions sent white-hot lava flowing over the land until it cooled and hardened into different kinds of igneous rock.

This building up of rock layers eventually buried the older material under thousands of feet of younger rocks. Eruptions and earth tremors subjected these rocks to varying amounts of heat and pressure, and this process decomposed the fossil material, changing it into new compounds composed only of the elements hydrogen and carbon. These substances are called hydrocarbons.

Because of the heat, the woody and leafy material of plant remains turned dark brown and then deep black. The great pressure of the rocks squeezed and compacted the material until it became as hard as rock. Today, men dig this black rock from the ground and call it coal.

In other parts of the ground, buried animal material was also being squeezed and heated. It chemically decomposed also, but formed hydrocarbons different from the vegetable matter. These were oily brown and black liquids that are called petroleum. The great heat further decomposed some of the largest hydrocarbon molecules into smaller ones. These tiny hydrocarbons, composed of only one carbon atom and four hydrogen atoms, make up the gas methane. Large amounts of methane formed along with the petroleum; so when we strike oil with drills, we usually find this natural gas along with it.

Coal, petroleum, and natural gas are called fossil fuels because they formed from the remains of those ancient living things.

Why Do Fossil Fuels Have Energy?

Fuels were once living material, and their energy was once food energy. Its source can be traced back to green plants, since all living things eat green plants, directly or indirectly. Green plants have a special ability to create their own food by a process called photosynthesis. Special cells in the leaves take carbon dioxide molecules out of the air and chemically combine them with water molecules sent up from the roots, to produce a molecule of food called glucose. At the same time, several molecules of oxygen are produced as left-over material, and this by-product is released into the air.

Additional energy is needed to make the glucose, and it is supplied by the sun in the form of light. A complex substance, chlorophyll, captures this light energy by using its outermost electrons to absorb it. In doing this, the plant has turned molecules with a low energy level into ones with a high energy level. The process of photosynthesis converts solar energy into chemical energy that is stored in the chemical bonds holding the glucose molecule together. Photosynthesis is called an endothermic, or energy-absorbing, reaction, because solar energy is needed to form products that are less stable than the reactants. This energy from the ancient sun, captured in food by green plants and passed on

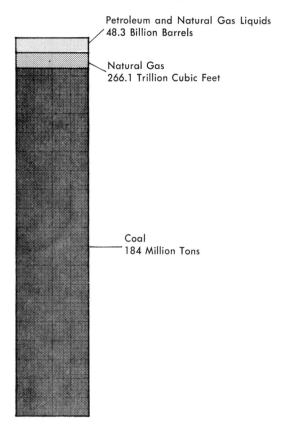

U.S. fossil fuel reserves.

to animals, ends up in our fossil fuels, from which we release it by burning.

What Happens When a Substance Burns?

The efficient use of fuels requires an understanding of the burning process. Only then can fuels be used with least waste and maximum output of heat energy. Burning is a chemical change in which the elements in a fuel combine with oxygen, with an accompanying release of heat energy. Chemists call this process oxidation. The burning process is the reverse of photosynthesis. It chemically combines the oxygen and glucose to re-form carbon dioxide and water. The reaction is exothermic, releasing the energy that was stored in the glucose chemical bonds. Although glucose is changed by the plant into many different molecules such as starch, cellulose, and protein, the energy remains within these bonds throughout the changes. Finally, when the organism dies, the energy can be preserved in the fossil material provided that bacteria do not have an opportunity to complete the decay process of slow oxidation.

Substances other than fuels can combine with oxygen. Iron combines to form iron oxide, or rust. Newspapers become yellow and crumble because the elements in the paper oxidize. These reactions occur slowly, but when a fuel burns, it combines with oxygen so quickly that noticeable heat and light are produced. Burning is called rapid oxidation.

Under What Conditions Will a Fuel Burn?

In order for burning to occur, atoms of a fuel and oxygen must combine. Since oxygen in the air and the elements in a fuel exist as molecules, they must first be broken down into atoms before the fuel can burn. This is what happens when the flame of a match is touched to paper or wood. The heat energy of the match flame causes molecules on the surface of the paper to move fast enough to break down into atoms, and then collide with oxygen and recombine to form carbon dioxide and water.

One necessary condition for burning, therefore, is that a fuel be heated to a sufficiently high temperature. This is called the kindling temperature, and it is not the same for all substances. Substances with a low kindling temperature are often used to start the burning of substances with higher kindling temperatures. A campfire is started by first burning paper; this in turn ignites small pieces of wood called kindling, which produces a fire hot enough to burn large logs. An ordinary friction match uses the same principle. The tip of the match is coated with a substance that has a very low kindling temperature. Merely rubbing the match on a rough surface creates enough heat to ignite it.

What Products Are Formed When a Fuel Burns?

Breathing on a glass mirror will cause water to form. This water, a compound of hydrogen and oxgen (H_2O), is produced by the oxidation of foods in the body. In a similar way, water is formed when a fuel burns. Another product of both respiration and burning is carbon dioxide.

The equation for the burning reaction may be written:

$$\text{Fuel} + \text{Oxygen } (O_2) \rightarrow \text{Carbon Dioxide } (CO_2) + \text{Water } (H_2O)$$

From this equation, can you determine the composition of a fuel? You are correct if you decide it must contain carbon and hydrogen, for the fuel is the only substance in the equation that could supply these elements. Thus, the common fuels (coal, oil, and natural gas) contain hydrocarbons; and the products of burning (water and carbon dioxide) are formed from the chemical combination of oxygen with these hydrocarbon compounds.

How Do Burning Fuels Cause Air Pollution?

The problem of air pollution is common to all of us. The black smoke often seen when fuels burn contains many particles of unburned carbon, or soot. That is the result of incomplete burning. The equation for this incomplete burning reaction is:

$$\text{Fuel} + \text{Oxygen} \rightarrow \text{Water} + \text{Carbon}$$

In a similar way, carbon monoxide rather than carbon dioxide may be produced when fuels burn. This highly poisonous gas is also formed by incomplete burning. The equation for this reaction is:

$$\text{Fuel} + \text{Oxygen} \rightarrow \text{Water} + \text{Carbon Monoxide}$$

In many cases, flame color is an indication of the completeness of burning. A flame produced during a complete burning reaction will have a blue color. If burning is incomplete, the flame color will be yellow or orange, this color resulting from the incandescent glow of hot, unburned carbon particles.

Incomplete burning is due to a lack of oxygen necessary to oxidize the fuel completely. Because oxidation is incomplete, much heat energy remains unreleased. These reactions may occur in automobile engines where the supply of oxygen is purposely limited, or in furnaces that are improperly adjusted or not correctly filled with the fuel.

Other qualities of the fuel may cause air pollution. For example, the element sulfur is always present in fossil fuels. When the fuel is burned, sulfur dioxide, an irritating and highly dangerous gas, is released. Air pollution also results when lead-containing compounds are added to gasoline in order to improve automobile engine performance. Lead poisoning, a very serious condition, may be caused by the release of lead compounds into the atmosphere.

How Are Fossil Fuels Prepared for Use?

Before fossil fuels can be used commercially, they must undergo several processes. First, they must be extracted from the earth. Then, each must be analyzed chemically in order to determine its composition and applicability. Many individual or groups of compounds with special uses may have to be separated out. In addition, the fossil fuels, which are often found in distant places, have to be transported to homes and factories many thousands of miles away. In the chapters to follow, we shall see how each of the three fossil fuels undergoes these processes.

QUESTIONS TO THINK ABOUT

1. What are the three most important fossil fuels?
2. What are some important ways we use fossil fuels?
3. When did the fossil fuels form?
4. What substances eventually became the fossil fuels?
5. How did sediments aid the formation of fossil fuels?
6. What is igneous rock?
7. What do we call the compounds in fossil fuels? Why?
8. Where did the energy of fossil fuels come from?
9. What is photosynthesis? What are its products?
10. What part does chlorophyll play in photosynthesis?
11. What is an endothermic reaction?
12. What occurs during burning?
13. What is oxidation?
14. What is the difference between slow oxidation and rapid oxidation?
15. What is an exothermic reaction?
16. What is meant by a fuel's kindling temperature?
17. What must happen to a fuel's molecules before it can burn?
18. Write an equation for the complete burning of a fossil fuel.
19. How do the products of burning prove that a fuel can be a hydrocarbon?
20. What are two undesirable products in incomplete burning?
21. What is the primary cause of incomplete burning?
22. Aside from air pollution, what is another undesirable result of incomplete burning?
23. Name two other pollutants that may be released during burning.

V

Coal And Coke

Coal is one of our very important materials. It is most valuable as a source of heat, and industry applies the energy of burning coal in many ways. It keeps buildings warm in winter. Its heat changes water to steam for generating electricity. Coal is also the source of energy for many of the industrial plants that produce steel, copper, aluminum, cement, and the countless items manufactured from these materials. By learning more about its composition, scientists have been able to use coal as a raw material for the production of plastics, explosives, drugs, and other valuable chemicals.

Are There Different Kinds of Coal?

There are three types of coal: *lignite, bituminous coal,* and *anthracite.* Each is the result of a different amount of heat and pressure having been exerted on it when it was buried deep in the ground. Nature's first step in the making of coal is the formation of peat. Peat is plentiful in swamps where dense vegetation is growing. It is the semidecayed material of plants that have died only recently and have not had time to be buried deeply. Brownish in appearance, peat looks more like wood than coal, and it is not truly a coal. Furthermore, it is not able to burn as hotly as coal because it contains considerable moisture that cannot be removed easily. In some parts of the world where other sources of heat are not available, peat is dug out of the swamps, dried in the air, and burned as a fuel.

Lignite, formed from peat, is considered a true type of coal. Brownish-black in appearance, it develops when peat is squeezed and heated enough by overlying rocks to decompose it and press it into rocklike hardness. Vast supplies of lignite are found in various parts of the United States, but little has been mined because it is of low quality as a fuel compared to bituminous coal and anthracite, which also exist in large deposits. Like peat, lignite contains a large percentage of moisture, which reduces its heating value.

The most abundant and economical fuel in this country is bituminous coal. Greater heat and pressure than lignite received were needed to produce this dark black material. It is one of our most important sources of heat energy, and, in addition, it is a storehouse of many important industrial chemicals. Experts estimate that our supply will last for several hundred years. Bituminous coal is often referred to as "soft coal," because when it burns, it becomes soft and plastic.

Anthracite, a shiny black coal, requires an enormous amount of heat and pressure for its formation. This necessary condition occurs only occasionally in the earth. It sometimes happens when deep rock layers containing coal seams are buckled and folded by gigantic forces deep in the earth, uplifting mountain ranges. The Appalachian Mountains in Pennsylvania were formed in this way, and this region holds our most abundant supply of both anthracite and bituminous coal. Anthracite does not soften when burning, and for this reason it is usually called "hard coal."

How Is Coal Mined?

Coal is plentiful in the United States, and mines exist in many sections of the country. Usually, the coal is found deep below the surface of the earth. In underground mining, deep shafts are dug down to the coal beds. Explosives are used to blast the coal loose, and special machinery is used to gather the coal chunks and bring them to the surface. Columns of coal several feet high are left standing in order to support the ceiling structure. To prevent cave-ins, special braces are installed to support the tunnel roofs

This giant continuous mining machine combines the functions of cutting, drilling, blasting, and loading. Carbide cutting bits literally rip coal out of the seam and gathering arms sweep the broken coal onto a conveyor, which discharges into shuttle cars. The overhead protective canopy provides strong protection for the operator in the event of a roof fall.

in potentially weak places. The mine is well ventilated by powerful fans that circulate fresh air through the tunnels. Damp powdered limestone is periodically sprayed through the tunnel in order to settle any dust in the air. Gas analyzers detect the presence of dangerous amounts of poisonous and explosive gases. In addition, all parts of the mine are frequently inspected to insure as little risk as possible.

Coal beds are also found at or near the earth's surface. This may be caused by internal pressure on the layers of rock and coal that causes them to be thrust up toward the surface. Erosion wears away the upper layers of rock, thus exposing the coal beds. In this situation, the more economical method of strip mining is used to

COURTESY AMERICAN ELECTRIC POWER CO.

Big Muskie, the world's largest dragline, removes overburden with a 220-cu.yd. bucket, large enough to hold twelve automobiles. The boom is longer than a football field.

extract the coal. Huge diesel-powered shovels lift the coal directly into trucks or freight cars for shipment to the user. However, even larger machines must first strip off an overburden of surface rocks and dirt many feet thick before the main coal body is reached. In any case, strip mining is much more efficient and economical than underground mining, as well as being less dangerous. Its one great disadvantage is that it can leave the scenic beauty of the countryside ruined with ugly, craterlike pits. Many states have passed regulations requiring the mining companies to landscape the exhausted mines and convert them into lakes and parks.

What Is the Chemical Composition of Coal?

The chemical nature of coal indicates that it formed from trees and other woody vegetation. Investigation by scientists shows that coal is made up of the two major elements found in all living things: carbon and hydrogen. The two are chemically combined to form the class of compounds called hydrocarbons. Since carbon and hydrogen can readily form hundreds of compounds by combining in varying proportions, coal is a complex mixture of hydrocarbons.

Here is an easy experiment you can do in order to understand something about the nature of the chemicals in coal. Light a candle. Did you ever wonder what causes the flame? Examine it closely. You can see that there appears to be something burning in addition to the wax or wick itself. Blow out the flame, and watch what comes out of the wick. Now you know what burns in a flame. It is a gas. To prove this, take another candle and light both of them, keeping them about a half-inch apart. Blow out one flame and quickly bring the lit candle to the rising wisp of smoke, but don't touch the wick itself. What happens? (The unlit candle catches fire.) You may have to repeat this a few times to be sure you really see that. A flame is actually a burning gas and, in this case, the gas acts like a fuse. Most fossil fuels burn with flames because they produce flammable gases when they are heated.

When a fuel is heated, some of the larger hydrocarbon molecules are broken apart by the heat energy and become gases that burn easily. This gaseous material is called volatile matter. For example, when wood is heated in a closed container, the larger molecules decompose into smaller, simpler molecules, some of which are gases, or volatile matter. When the volatile matter in the wood is exhausted, the flame dies out and the remaining combustible material just glows as it burns. This material is carbon, sometimes called fixed carbon because it does not vaporize easily. Thus, when a fuel is heated, it breaks down into volatile matter and carbon.

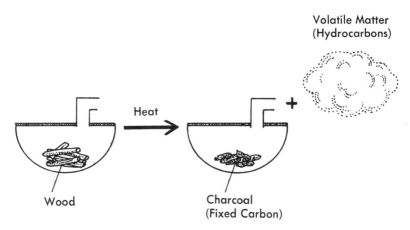

Volatile Matter
(Hydrocarbons)

Heat

Wood

Charcoal
(Fixed Carbon)

The destructive distillation of wood.

The amount of volatile matter in coal determines its type. Coal that is subjected to greater amounts of heat and pressure in the earth has a lower amount of volatile matter, whereas the percentage of fixed carbon remaining in it increases. Wood, peat, and lignite give off a greater amount of volatile matter when heated. Anthracite gives off hardly any, and bituminous varies between these two extremes.

Why Are Anthracite and Bituminous Coals Most Often Used?

Anthracite and bituminous coals are the two most important solid fuels. Anthracite contains about five percent volatile matter.

If a small lump of anthracite is held in a flame, it will not readily ignite because there is so little volatile matter. The fixed carbon takes longer to begin burning than the gas, but once started, it burns steadily and completely, with little smoke and flame. Homeowners who still use coal prefer anthracite because inexpensive furnaces can burn it completely, with little attention being required. In addition, it burns cleanly, causing only a minimum of soot and other air-polluting substances. Because of this, coal-burning industries located in cities are considering the use of anthracite to replace bituminous coal in their furnaces. However, they will have to pay a higher price for anthracite, because it is more expensive to mine and is in shorter supply than soft coal.

Bituminous coal usually contains a lower percentage of fixed carbon and a higher percentage of volatile matter than anthracite. If bituminous coal is held in a flame, it quickly begins to burn, giving off a long yellow flame because of the fairly high percentage of volatile matter. The large amount of smoke and soot (carbon) given off show how difficult it is to burn completely. Combustion engineers have designed large furnaces with huge blower fans that force in air to oxidize the coal more thoroughly and minimize smoke and soot. Other devices connected to the chimney, called electrostatic precipitators, remove smoke particles by attracting them to electrically charged plates, much as dust is attracted to a phonograph record. Such equipment is expected to minimize air pollution so that bituminous coal may continue to hold its position as the most widely used solid fuel.

How Can Coal Be Improved?

The burning characteristics of bituminous coal can be improved by changing it to coke, using a process called destructive distillation. The result of this is similar to what happens when you heat wood in a closed container so that it decomposes without burning. In destructive distillation, pulverized bituminous coal is placed in

The coking process. Approximately 13 tons of coke pour from one of the coke ovens into a waiting railroad car at the Sparrows Point plant of Bethlehem Steel during a coke push. The conversion of soft coal to coke takes from 18 to 20 hours.

closed, air-tight ovens and heated until the volatile matter is completely driven off. The fixed carbon remaining in the oven is the coke. The volatile matter is not wasted, but is collected and separated into valuable chemicals used to make explosives, fertilizers, drugs, dyes, plastics, and other products of the chemical industry.

The cooled coke is a light, hard, porous mass. The particles of coal have partially melted and fused together while the escaped volatile matter has caused thousands of tiny holes to form in the remaining material. This porosity allows air to circulate within the coke, enabling it to burn rapidly and evenly. It also enables the fuel to be burned completely, eliminating most of the smoke and soot produced when soft coal is burned. Thus, coke resembles anthracite in the way it burns with the added advantage that its porous structure causes it to burn faster and reach higher temperatures.

As a result of these advantages, and also because it remains hard and strong when it burns, coke is an excellent fuel for iron-making furnaces. The iron and steel industry uses millions of tons of coke each year. In their giant furnaces, the coke supports the weight of several thousand tons of raw material without weakening as it burns because, like anthracite coal, it does not soften when hot.

What Is Coal's Future?

Coal represents our most abundant fossil fuel, and the United States Bureau of Mines estimates our reserves at 3 trillion tons —enough to last several hundred years. Even though the use of coal has been limited because of the inconvenience of shipping, handling, storing and burning a solid fuel, its abundance assures it a growing and permanent place in the economic and technical life of the world.

In power plants built right at the site of the mine, coal is burned for generating electricity by steam. In addition, as our liquid and gaseous fossil fuels are consumed because of the tremendous demand for them in transportation, industrial, and household needs, they will be replaced by similar hydrocarbon substances made by chemists from coal. Finally, coal will become an increasingly important source of raw materials for the chemical industry, which now uses it only to the extent of obtaining by-products from the coking process. Thus, coal, one of man's oldest fuels, will play an even greater role in the world of the future.

QUESTIONS TO THINK ABOUT

1. What is the primary use of coal? *to heat*
2. Name at least five products that are produced using coal as either a fuel or a raw material.
3. What is peat? Why is it not a good fuel?
4. How does lignite develop?
5. Compare bituminous coal and anthracite with respect to: (a) appearance; (b) origin; (c) abundance.
6. Why are bituminous coal and anthracite called soft coal and hard coal respectively?
7. Describe two types of coal-mining operations.
8. What dangers are involved in underground mining?
9. In what way is strip mining more advantageous than underground mining?
10. What is a possible disadvantage of strip mining?
11. What is a flame?
12. What is volatile matter?
13. Compare bituminous coal and anthracite with respect to: (a) volatile matter; (b) fixed carbon; (c) ease of burning; (d) cost; (e) pollution problems.
14. Describe the destructive distillation process.
15. What is coke?
16. What are the properties of coke?
17. What are some products that can be made from the volatile matter?
18. Why will coal become increasingly important?
19. How will coal be used in the future?

VI

Petroleum and Natural Gas

Every day in the United States, more than eighty million automobiles and trucks carry people and products over our nation's highways. In addition, thousands of airplanes cross our skies, while passenger ships and freighters of all kinds sail over the world's oceans. The energy needed for each of these forms of transportation is produced by the burning of products that have been developed from petroleum.

During the winter, most of our apartment houses and private homes are heated by burning fuel oil or gas. Even in the summer, we must burn some fuel to make hot water for washing and bathing. Most kitchens have stoves that burn natural gas for cooking. In addition to serving as fuels, petroleum products are used for many industrial products. Lubricants, waxes, preservatives, and solvents are among the more important of these items.

In the sections to come, we shall discuss petroleum and natural gas: how they are found, brought to the surface, and processed into the many products that make up a multibillion-dollar industry that employs more than two million people in the United States.

How Does an Oil Field Form?

Fossil fuels were formed by the partial decay of dead plants and animals. The remains were buried by sediments of sand, clay, and

61

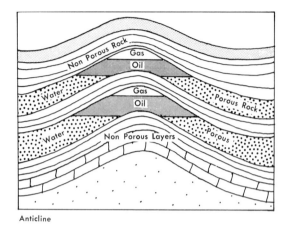

Anticline

Fault

Stratigraphic Trap

Geologic traps.

lime, which later hardened into sedimentary rock. Each sediment formed its own kind of rock: sandstone from sand, shale from clay, and limestone from lime.

In the case of petroleum, most of this process occurred under the sea, where fine grains of minerals were packed together. The oil and natural gas forming in this material, along with salt water, collected in the porous spaces between the coarser particles of sandstone. Underground pressure squeezed oil from porous layer to porous layer toward the surface of the earth. Sometimes, non-porous layers of stone formed geologic traps that blocked the upward movement of the petroleum, making it accumulate until an underground oil pool formed.

All the great oil fields have formed from one of three types of geologic traps. They are the anticline, the fault, and the strati-graphic trap. Although each one begins as a series of layers of porous and nonporous sedimentary rock, they develop by different processes.

The anticline develops when the crust of the earth is folded by internal pressures, forcing the layers of rock into dome-shaped structures that extend for several miles through the ground. The petroleum collects under the top of a dome of dense nonporous rock. Anticlines are the largest and most important of the geologic traps.

The fault trap forms when internal pressure causes the layers to crack instead of fold. The pressure then lifts one side of the fault at an angle to the other, and the petroleum collects in the upper part of the angle.

Stratigraphic traps form from the remains of ancient seashores. In time, the beaches are covered by clay that hardens into dense shale and traps the marine animals and plants that form petroleum within the beach sand. The petroleum collects in the pinched-off layer of sand that once was the beach.

How Are Oil Fields Found?

The job of the prospector is to find these geologic traps, even when they are deep underground. Today, the demand for pe-troleum is so great that the scientists' search extends even into the ocean waters. These rugged people also brave the cold of the arctic regions and the heat of the deserts and jungles in their adventurous quest for oil.

The first oil fields were found by chance in locations where the petroleum or natural gas was forced to the surface by the pressure inside the earth's crust. The oil was able to escape the traps

COURTESY EXXON CORPORATION

An offshore mobile rig used in the North Sea. It can operate in water depths of 600 feet. This rig struck oil off the east coast of Scotland.

because earthquakes probably cracked the nonporous layers, creating a path outward and upward. These oil seeps were the places where the early oilmen drilled their wells, the first of them at Titusville, Pennsylvania, in 1859.

The increased demand for oil soon drained these early, easily found fields. Wildcatters, oilmen who gambled their fortunes, began drilling wells at any site that seemed similar to a region where oil had already been found. Like most gamblers, the majority were soon bankrupt, for there is very little likelihood of striking oil by pure chance. Today, petroleum geologists using scientific methods have greatly improved these chances.

Geologists look for oil by seeking the underground traps in which it collects. Measuring the depth and size of underground rock layers, they drill in areas that indicate the presence of these traps. Since most of them are dry, either because petroleum never formed in the area or because it escaped through cracks, it is only the drill that can tell the oilmen if their search has been successful.

Even with the most modern equipment, only one well out of nine strikes petroleum in newly mapped areas. Without scientific methods, the rate of failure would be much higher.

The scientists begin their search by using aerial or satellite photography to map areas of the earth's surface covering hundreds of square miles. The purpose is to find surface features such as low hills, gentle mounds, and long rifts that might indicate the presence of anticline or fault traps.

In aerial mapping, the geologists also require the use of gravimeters and magnetometers. These devices measure the force and direction of the land's gravitational and magnetic fields, which are affected by the types of rock in an area. The variations discovered by the instruments give the scientists more clues about the types and locations of rock layers below the surface.

The scientists then travel by truck, boat, or helicopter to the regions that may contain geologic traps. They survey and map in detail the different kinds of rock layers they find. Specially trained geophysicists use sound waves to locate and pinpoint the different types of rocks that form the sedimentary layers. The sound is produced by explosions set off in wells drilled about 70 feet deep. The sound waves travel down into the rock, and the rock layers reflect part of the sound back to the surface like an echo. The geophysicist uses a seismograph, which can sense the vibration of the ground, to record and measure the sound impulses reflected.

The geophysicist analyzes these measurements and then constructs maps of the underground rock layers. These maps show whether or not a geologic trap exists that may contain oil. If the presence of a trap is indicated, the next and most crucial step is drilling down through the rock layers to the trap region to see if it actually contains petroleum or natural gas. As the oilmen say, "Only the drill can find the oil."

How Is Oil Brought to the Surface?

After the geologists, geophysicists, and surveyors have located possible geologic traps, the drilling crew moves into the area with heavy equipment. Many times, this means being airlifted by helicopter to some isolated spot in rugged country. In other places, the drilling crews work from platforms set up over shallow waters often several miles offshore.

In soft earth where the oil-bearing formations are close to the surface, a simple and inexpensive method known as cable tool or percussion drilling can be used. This method, based on the principle of the pile driver, punches a hole in the earth.

The usual procedure for drilling wells deep into the ground or through hard layers of rock is the rotary drilling technique. This process is the most efficient way to drill the deep wells needed today to find petroleum deposits. It is based on the idea of boring holes into wood with a hand or power drill. The faster speed of penetration of rotary drilling compared to the percussion method makes it a more economical way to drill deep wells, despite the need for more expensive equipment.

The first task of the drill crew is to erect a derrick, the familiar tower of steel scaffolding that can make an oil field look like a man-made forest. The derrick is used to hoist the many sections of pipe and casing to be used in drilling the well. Then the drill

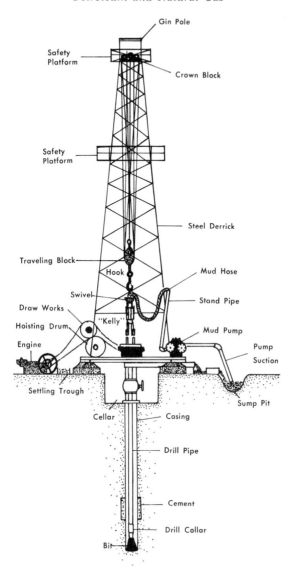

A rotary drilling rig.

crew sets up the power plant and connects it to the turntable, which will actually rotate the drill bit. A square hole is cut in the center of the turntable, and through this is fitted a hollow square pipe called a "kelly." The kelly in turn is fitted to a hollow pipe section called the "drill stem." Fastened to its end is the drill bit, the part that does the actual grinding and boring into the earth. Drill bits are made of various hard materials and come in different shapes depending on the nature of the rock being pierced.

As the bit cuts deeper into the ground, new sections of pipe are added to the drill stem to increase its length, and the derrick tower aids in this process. When the bit is worn and must be replaced, the whole drill stem must be pulled out and uncoupled. This is a tedious, expensive job that can take more than two days if the well is deep. After a new bit is attached, the whole drill stem is reassembled and lowered back into the hole.

Aiding the drilling process is a special mud that is pumped down the well through the kelly and drill stem to squirt out around the drill bit. The mud cools the bit and lubricates it, making the grinding process easier. The mud then rises up the well around the outside of the drill stem. On its upward path, it carries the drill cuttings out; its pressure prevents cave-ins of the well sides; it plasters the sides of the hole to prevent water seepage.

The drill crew constantly samples the cuttings that the mud brings up, examining them for signs of oil. Certain fossils are one clue that oil may be present. In addition, special equipment is lowered down the well to "log" electrically or detect the presence of oil in the penetrated rock layers. The oil reveals its presence by affecting the resistance of the rock to electric currents. If the reports indicate that recovery chances are good, a gun perforator is lowered to the bottom and fired, shooting bullets out into the surrounding rock to allow the oil to enter the well more easily.

If the oil is under great pressure, steps must be taken to control its rise to the surface. A system of control valves called a "christmas tree" is fitted to the casing at the top of the well. The oil is then allowed to rise up the well and is directed by the valves to storage tanks. Under this modern system, the once-familiar "gusher" of oil that shot into the air like a geyser when a well came in has become a thing of the past.

Sometimes the natural pressure in the oil formation fails. Then the oilmen must attach pumps to the well to bring up the oil. In order to conserve this valuable resource, the engineers have developed other methods to force the oil out of even the finest pores in the underground rocks and bring it to the surface. Sometimes they pump water down a well under great pressure, forcing it into the porous rocks below. Sometimes they pump down air and flammable gases, which they ignite below ground. The purpose in both cases is to renew underground pressure that will force the last remaining oil up other wells to the surface.

What Is the Chemical Nature of Petroleum?

Have you ever seen films showing a gusher of crude oil shooting into the air? This black liquid is perhaps the most important material of our time. Countries throughout the world want to acquire as much oil as possible, and a good deal of recent history may be told in terms of their attempts to do so.

It takes knowledge of science and technology to make oil useful. Every drop must be carefully treated in order to extract all the valuable material it contains. Petroleum is so complex in composition and so rich in possible products that scientists have not fully learned about its chemical nature. Although thousands of products have been created, thousands more remain to be discovered.

In order to utilize any material, its properties must be understood. Since the properties of petroleum are mainly determined by its chemical structure, the first step in exploiting it must be to understand its composition.

Chemical analysis shows that petroleum is a mixture of hydrocarbon compounds, and each has its own characteristic chemical structure. Some are very small, made of only five atoms. Others are very large, containing hundreds of atoms. In addition, chemical combinations take many different shapes. Branchlike structures may form. Other molecules are shaped like rings with five or six sides. Complex molecules having rings combined with branches can also occur. These varied shapes give hydrocarbons many important properties.

What Effect Does Chemical Structure Have upon Petroleum Properties?

The payoff of all scientific investigation occurs when the knowledge gained is applied. Understanding the chemical structure of the hydrocarbons in petroleum enables us to understand their properties. Once the properties of the individual compounds are understood, we can develop efficient ways to separate and make use of them.

After many years of experimenting, chemists learned the general characteristics of these compounds. The major discovery was the relationship between the size of a hydrocarbon and its physical appearance. Can you guess what the difference in appearance would be between one liquid composed of large molecules and another composed of smaller molecules? If we separate petroleum into these two portions, we find that the larger molecules form a viscous, or thick and syrupy, liquid, whereas the smaller molecules form a liquid with low viscosity, which can flow more quickly than water. Scientists explain this difference by saying that the forces of attraction between the large, heavy molecules are greater than those between the small, lighter ones. This is proven by the fact that the very smallest are actually gases. The forces of attraction in these molecules are so weak that they do not remain together as do those in the liquid.

Can you figure out what effect molecular size would have on the boiling point of these compounds? You are correct if you think that an increase in size would cause an increase in the boiling point, because as attractive forces between the molecules increase, the amount of kinetic energy required to overcome these forces also increases. Since kinetic energy is increased by raising the temperature, the larger molecules have higher boiling points. The larger molecules also condense sooner when the temperature is lowered. Thus, they have a higher condensing point than smaller

THE BOILING POINTS OF HYDROCARBONS

Name	Formula	Structure	Boiling Point
Methane	CH_4		$-161.4°C.$
ISO-Octane	C_8H_{18}		$99.3°C.$
Dodecane	$C_{12}H_{26}$		$214°C.$

molecules. Actually, the boiling point and the condensing point of a compound are at the same temperature.

How Are Petroleum Compounds Separated?

The chemist tries to use these properties as a basis for separating the hydrocarbons. Can you see how they may be separated because of differences in their boiling-condensing points? Imagine that you have a flask containing petroleum. How could you separate the various compounds?

To give you a clue, you might try this experiment. Pour equal amounts (about one teaspoonful) of rubbing alcohol, water, and mineral or machine oil into three separate dishes. Place these near a warm oven or radiator and observe which is first to evaporate. What must be true about the boiling-condensing point of the oil compared to the other materials? If the three liquids were mixed together in one dish, which would be the first to escape as a vapor? You are right if you realize that the one with the lowest boiling-condensing point is the first to evaporate. Petroleum scientists have adapted this principle by first boiling the crude oil and then condensing the vapors at controlled temperatures to separate the mixture. This process of vaporizing and condensing is called distillation.

Petroleum is composed of hundreds of hydrocarbons, and, although each has a different boiling point, many of them are only a fraction of a degree apart. The separation of petroleum into individual compounds is, therefore, extremely difficult to accomplish. In actual practice, compounds with similar boiling points vaporize and condense together when they are distilled. The compounds in each group are called fractions. If necessary, the hydrocarbons in each fraction may be further distilled into smaller groups, but the fractions themselves may be of such use that further refinement is unnecessary. The process by which these separations are carried out is known as fractional distillation.

How Is Fractional Distillation Accomplished?

Several attempts have been made by industry to provide a chemical plant that would perform fractional distillation quickly and economically. Over the years, only one type of structure has stood the test of time: the fractionating column. The fractionating column represents one of the most outstanding achievements of

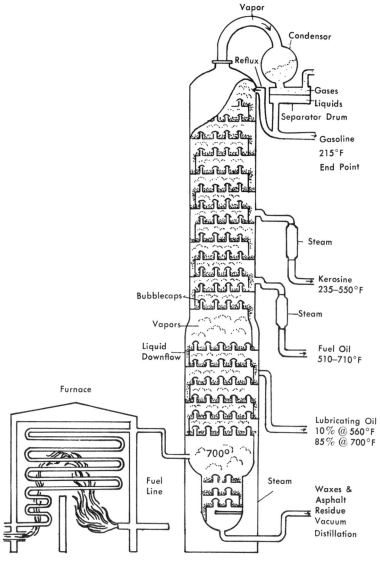

Vapor

Condensor

Reflux

Gases
Liquids
Separator Drum

Gasoline
215°F
End Point

Steam

Kerosine
235–550°F

Bubblecaps

Steam

Vapors

Fuel Oil
510–710°F

Liquid
Downflow

Furnace

Lubricating Oil
10% @ 560°F
85% @ 700°F

700°

Fuel
Line

Steam

Waxes &
Asphalt
Residue
Vacuum
Distillation

A fractionating column.

modern technology. Towering high above the ground, this structure converts crude oil into several useful products in one major operation.

Despite the many complex chemical and mathematical calculations required for its design, once completed, the column is simple in form and its basic method of operation is easy to understand. Shaped like a gigantic cylinder, the tower is made of metals such as stainless steel or aluminum, which will not corrode when in contact with hot crude oil. Round trays, spaced at various levels, are welded to the inner sides of the cylinder. Several holes are drilled in the trays so that crude oil vapors can pass from one level to another. Pipes exiting from the tower at various trays carry out liquid fractions that condense there.

Crude oil is first led to a furnace, which vaporizes most of it. The vapor is then piped into the tower near the bottom trays, where it begins to cool as it rises. When the vapor has cooled sufficiently, it condenses and collects in the tray at that particular condensing point.

In general, the size of the molecule determines the level and the tray at which it will condense. Larger molecules condense first in the tower and fall toward the bottom. Vapors of the smaller hydrocarbon molecules rise upward through the holes in the trays to the colder parts of the tower, where they too finally condense. Each hole is covered by a bubble cap, which traps the rising vapors and forces them to bubble through the liquid that has already condensed on its tray. Such close contact between liquid and vapor assures condensation of the correct molecules on each tray. The very smallest molecules do not reach temperatures cold enough to condense them even at the highest level; therefore, they are piped out as a gas from the top of the tower.

As you can imagine, one of the most delicate operations of this process is maintaining a constant temperature at each tray. If this temperature varied, the composition of the condensing vapors would vary as well, and the fractions would contain many types of molecules. Since hot vapors are constantly entering the tower, the temperature at each tray naturally has a tendency to rise. The engineer at the refinery counteracts this by liquefying part of the gas that has been piped out at the top of the tower in a water-cooled condenser. A calculated portion of this cold liquid is then sent back into the tower at the top, where it falls onto the highest

The oil refinery is one of man's greatest achievements. From it come many thousands of products needed for modern living.

tray. This cools the liquid already on that tray just enough to balance its tendency to heat up. Overflow pipes are provided on each tray to carry similar cooler liquids to the next lower one. As the cold liquid flows down, the level on each tray is raised, and some liquid flows through the overflow pipe down to the next lower level. Since this liquid is cooler than the liquid below, the lower tray cools and the entire process repeats itself. Thus, on each tray, the temperature remains steady as cold liquid from above counteracts rising hot vapors.

What Are the Major Crude-Oil Fractions?

As the distillation continues, each fraction is drawn off from its tray. Thick greases and asphalt exit from the bottom of the column. Farther up, lubricating oils are piped out. Next, lighter oils used in heating homes are removed. Kerosene flows from trays in the upper portion of the tower. Gasoline and aviation fuels lie on the uppermost trays. Finally, gases that do not condense at the top or in the condenser are led away from the column.

The number of trays in a tower may vary, and the petroleum may therefore be broken down into a few crude fractions or several purer ones. In most cases, the fractions must be further refined by additional distillations to yield more usable products.

Asphalt. The material that lies at the very bottom of the tower is known as *asphalt.* Dark in appearance, it has molecules so large that it appears as a solid at room temperature; however, at higher temperatures, asphalt softens and flows. Because of this property, it is applied mainly as a surface coating.

By far the greatest part of asphalt produced is used in the construction of roads. After being mixed with gravel at a high temperature, it is applied to streets and highways. Its plasticity enables this mixture to be flattened by steamrollers. When cool, it hardens and binds the gravel together, forming a strong surface. The molecules in asphalt repel molecules of water, thus waterproofing the road and causing faster drying after rainfalls. In addition, asphalt diminishes automobile noises and vibrations, making for a safer and smoother ride. Other applications that take advantage of these properties are in roofing construction, paints, corrosion-resistant coatings, and packaging.

Waxes. From the bottom trays of the tower emerge waxes, which have somewhat lower condensing points than asphalt. Following their separation from the crude oil, these fractions are further refined, and they condense above room temperature to form solids containing mainly very long straight chains of hydrocarbons.

The outstanding properties of waxes are the ease with which they can be softened, shaped, and then rehardened; their ability to vaporize when heated; and the ability of this vapor to burn. Waxed paper, so useful in keeping bread and other foods fresh, is a very important product. Place a drop of water on a piece of waxed paper. Notice how the water does not soak into the paper. If it is tilted, you will see that the water will easily run off and leave very little trace of wetness. This ability of wax molecules to repel molecules of water is put to use in the food-packaging industry, where paper and cardboard are wax-coated to make them moistureproof. Another important use of waxes is in the manufacture of candles. After the wick is lit, part of the wax near the wick vaporizes. The vaporized wax then burns, causing more wax to vaporize, and the process continues until all the wax is vaporized and burned.

Lubricating Oils. These oils are among the fractions having high boiling points found in the bottom portion of the tower. They are mainly used on the moving parts of machines such as engines and gears and form a layer between the parts in order to reduce the amount of wear and corrosion. You can easily understand the effect of lubricating oils by first rubbing your finger hard on any flat surface and then doing the same after placing a few drops of oil on the finger.

In machinery, the oil used must be able to lubricate properly to reduce wear, but must not be so thick that it restricts the movement of the parts. Each oil is therefore classified according to its viscosity—the difficulty with which it flows—in order to determine under what conditions it is to be used. For automobiles, the Society of Automotive Engineers (SAE) has established a numerical system to classify engine oils. For example, SAE 40 oil is a thick, viscous liquid, whereas SAE 10 is much thinner. Since the viscosity of most oils decreases as the temperature increases, in parts of the country where temperatures vary widely from season to season, different oils must be used at different times so that lubrication performance remains good. Oils with low SAE numbers are used in winter, and oils with high SAE numbers are used in summer.

Fuel Oils. Liquid fractions with lower boiling points and viscosities than lubricating oils are next to be drawn from the tower. These oils flow and vaporize easily enough so that they may be used as fuels in homes, factories, and heavy-duty engines. Of course, their properties vary according to the sizes of their molecules. For example, Number 1 fuel oil, composed mainly of smaller molecules, flows easily and has a low boiling point, whereas Number 6 oil flows and vaporizes with difficulty.

They also differ in the amount of heat they produce when burned. Heavier fuel oils may produce close to 10 percent more heat than lighter ones. As a homeowner, you may someday have to choose which type of oil to use in heating your home. Although Numbers 5 and 6 oil produce the most heat, expensive equipment may have to be used in order to keep the viscous oil flowing steadily and to vaporize it so that complete and efficient combustion takes place. Many electric companies formerly used the less expensive high-numbered oils to generate electricity; however, many have now decided to use oils of lower viscosity in order to reduce the air pollution caused by inefficient burning.

Kerosene. Kerosene lies in the upper portion of the tower, above the fuel-oil fractions. Since it is less viscous and more easily vaporized than heating oil, kerosene found great use as a fuel in lamps to illuminate homes and streets in the last half of the 19th century. It was the most important petroleum product at that time. Although it has been replaced by the electric light in the United States, many parts of the world still use it for lighting purposes. In recent years, kerosene has again grown in importance because of its increasing use as a fuel in jet airplanes.

Gasoline. Gasoline is produced near the top of the tower. Gasoline fractions are mainly used as a fuel for automobiles and airplanes. The molecules in these fractions are small, containing as few as five carbon atoms. They vaporize easily, burn quickly, and supply rapid power to engines.

In order to meet varying conditions, gasoline fractions are blended so that they may have suitable properties. Among the most important of these properties is the knocking quality of the gasoline, which measures its efficiency of combustion in the engine. Knocks are caused by vibrations in an engine when the combustion of gasoline vapor is not smoothly accomplished; this causes the engine to lose some of its power. Each blend of gasoline may be given an octane rating according to the smoothness with which it burns. The knocking quality of the gasoline being tested is compared with that of various mixtures of two hydrocarbons, iso-octane and heptane. The gasoline is then assigned a number corresponding to the percentage of iso-octane in the mixture it most closely resembles. For example, a gasoline with an octane rating of 100 would have knocking qualities equal to 100 percent iso-octane, and a gasoline with an octane rating of 90 would have knocking qualities equal to a mixture of 90 percent iso-octane and 10 percent heptane. Since iso-octane knocks very little and heptane knocks very frequently, the better grades of gasoline have high octane numbers. In order to improve the octane rating of gasoline, small amounts of tetraethyl lead are often added. Next time you are in a gasoline station, look for a sign on the pump reading "contains lead." In recent times, companies have made more gasolines with no lead compound added, because lead is poisonous and pollutes the air. Most automobile gasolines have octane numbers ranging from 75 to 100, and aviation fuels have octane numbers above 100 (they knock less than 100 percent iso-octane).

Natural Gas. The smallest hydrocarbon molecules emerge from the fractionating column as a gas. When underground, natural gas either forms above the liquid crude oil or is dissolved in it under pressure. When pressure is removed, the gas comes out of solution and can be piped away. In addition, dissolved gas is separated from the liquid in the distillation column itself. The major compound in natural gas is methane.

Natural gas is easy to ignite and burn completely with the simplest of equipment. Unlike petroleum and coal, gas offers no storage problem; most gas is delivered to the consumer by pipe, and the producing company takes care of the storage in huge underground reservoirs. Small-volume consumers in isolated areas receive their fuel in the form of liquefied petroleum gas (LPG) stored in small pressurized tanks. The LPG consists of slightly heavier hydrocarbons, propane and butane, which liquefy more readily than methane under moderate pressure.

Natural gas is used as a fuel for heating homes and factories. You probably use natural gas for cooking. In addition, natural gas may be further separated into individual hydrocarbons, which can be stored and shipped in a manner similar to LPG. Propane may be used in torches for melting metal and welding. Butane gas, long used for cooking and heating, has become popular as a fuel in cigarette lighters. The cost of natural gas is greater per Btu of heat delivered than other fossil fuels, but its cleanliness, ease of control, and high heating value make it desirable for many industries.

Can the Usefulness of Petroleum Be Increased?

At the time petroleum was first processed in large amounts, kerosene for light and heat was a great necessity of most people. Fortunately, petroleum contained quantities of kerosene that were enough to meet the demand. Since automobiles and airplanes had not yet been invented, not much gasoline was needed, and petroleum contained enough to satisfy our requirements. Although the times have changed, petroleum has not. The composition of petroleum that comes out of the earth today remains the same as it was one hundred years ago. As time passed, the crude oil was in danger of becoming less useful. While the demand for kerosene decreased because of the increasing use of electricity for light, the refinery was extracting more than could be used. At the same time, as the need for gasoline increased, the amount the distillation column supplied could not fill our requirements. Imagine that you were a chemist or petroleum engineer at the time the changes were taking place. How would you have tried to solve this problem?

The answer lies in our ability to chemically convert the excess kerosene to gasoline. This is accomplished by the process known as cracking, and its name alone should give you a hint as to how it works. You have already seen that kerosene molecules are larger than those of gasoline. In cracking, the kerosene molecules are split into those of gasoline. Other large molecules may be cracked into gasoline as well. In another process, gasoline molecules may be made by chemically combining the smaller molecules that leave the top of the tower. This process is known as polymerization.

In order to speed up polymerization and cracking reactions, heat and pressure must be applied to the reacting materials. Catalysts are also used to increase the rate of the reaction. These are materials that interact with the hydrocarbons so that less energy is needed when the actual cracking or polymerization takes place. The catalyst itself remains behind at the end of the reaction in an unchanged condition.

Petroleum is now being used as a starting material for manufacturing thousands of other products, and many look nothing like the original fractions from which they are made. Among these materials are synthetic rubber, plastics, dyes, paints, cosmetics, and medicines. Obtaining so many products from crude oil is one of the great wonders and satisfactions of modern technology. It is probably the best example of our ability to unlock and use the secrets of nature. As you can well imagine, hundreds of people have devoted their lives to this work and are constantly seeking better methods of production. Their research has rewarded us well.

QUESTIONS TO THINK ABOUT

1. What materials form petroleum?
2. What geologic events cause an oil pool to form?
3. How does an anticline trap develop?
4. How does a fault trap develop?
5. How does a stratigraphic trap develop?
6. What is the function of an oil trap?
7. What is a wildcat well?
8. What is the chance of striking oil in a newly mapped area?
9. What is the purpose of gravimeters and magnetometers?
10. How do scientists use a seismograph?
11. How can we determine whether oil is present underground?
12. How does percussion drilling work?
13. When is rotary drilling used?
14. Describe the purpose of the: (a) derrick; (b) kelly; (c) bit.
15. Why does the drilling crew use mud?
16. What methods are used to test for oil as the drilling proceeds?
17. Why are there no more gushers?
18. What steps can be taken to correct low pressure in the well?
19. Describe the size and shape of petroleum molecules.
20. What effect does molecular size have upon viscosity?
21. What effect does molecular size have upon boiling point?
22. What is the relationship between boiling point and condensing point?
23. Define distillation.
24. Why is the distillation of petroleum called fractional distillation?
25. Describe the fractionating column.
26. How does crude oil enter the column?
27. What happens to the crude oil when it enters the column?
28. What determines where the molecules will condense?
29. Where do the large and the small molecules condense?
30. What happens to the smallest molecules?

31. What is the purpose of a bubble cap?
32. What purpose does the condenser serve?
33. Why does the column have overflow pipes?
34. Name the major crude-oil fractions, and state in which part of the column they condense.
35. Name three uses of asphalt.
36. What properties make wax useful?
37. Define viscosity.
38. How does the temperature affect viscosity?
39. Compare Number 1 and Number 6 fuel oil.
40. Name an advantage and a disadvantage of using a high-numbered fuel oil.
41. How is kerosene used?
42. What does octane rating measure?
43. What is an advantage and a disadvantage of adding tetraethyl lead to gasoline?
44. What are three compounds found in natural gas?
45. Why is LPG made?
46. What are some uses of natural gas?
47. What does cracking do?
48. What does polymerization do?
49. What is a catalyst?
50. What other important materials can be made from petroleum?

VII

Nuclear Fuels

In preceding chapters, we saw how the chemical energy used to join together the atoms in a fossil fuel molecule is released when burning occurs. Most of the energy we use is released in this way. However, the nuclear forces binding the subatomic particles (protons and neutrons) together in each atom's nucleus are far greater than that.

How Were the Secrets of Atomic Energy Discovered?

Study of atomic energy began in earnest in 1896 when the French scientist Henri Becquerel discovered that powerful invisible rays were being emitted from certain minerals. Such emanations were the result of radioactivity. These rays were similar to light rays because they were able to fog photographic film. To the amazement of the scientists of that era, energy had come from lifeless rock—and without burning.

By the early part of the 20th century, it had been established that radioactivity was accompanied by changes in the atoms themselves. Elements were being destroyed, and new ones were forming to replace them: a process called transmutation. This was a great shock to scientists at that time because it had been believed that atoms were indestructible. Surprising, too, was the fact that the

85

total mass involved in the process had decreased slightly. In 1905 the German-born physicist Albert Einstein proposed an equation that related this loss of mass to the energy emitted by the radioactive material. The equation, $E = mc^2$, showed that when a minute amount of matter is converted a tremendous amount of energy is released. Einstein's theory of relativity, as it was called, had little practical importance until, during the 1930's, methods were developed to cause transmutations that had the potential to release energy in staggering quantities.

During the early 20th century, much research was done to determine the nature of the particles that make up the atom. The British physicist Ernest Rutherford made use of alpha particles, which are emitted by radioactive elements, to bombard atoms of nitrogen. He discovered that a positively charged subatomic particle was released, and named it the proton. Both protons and alpha particles were then made to collide with the atoms of many elements. The process was aided by the development of machines such as the cyclotron and the betatron, which accelerated the particles, giving them greater energy and making collisions more effective. The transmutation of various elements was caused by their collisions with the high-energy particles in these machines.

A major drawback to using both protons and alpha particles is that they are positively charged, and are therefore strongly repelled by the similarly charged nuclei with which they are supposed to collide. In 1932 the British physicist James Chadwick identified the neutron as a subatomic particle. Because it had no charge, the neutron could collide much more easily with atomic nuclei. The Italian physicist Enrico Fermi used neutrons extensively to bombard atoms. At times, the neutron would be "captured" by a nucleus, causing an increase in atomic mass. Sometimes beta rays (electrons) would be released. Occasionally, the newly formed nucleus would become so unstable that it would split violently to form atoms of much lighter elements. At that time, some mass would be lost and energy released. This disintegration process was most intensely investigated by the German physicists Otto Frisch and Lise Meitner. They called it nuclear fission.

Although only a single neutron is needed to begin a fission reaction, up to three neutrons are released from each splitting atom during the process. This implies that a chain reaction might occur if these neutrons were to initiate other collisions to continue the process. If, for example, two neutrons were released for each collision, the number of free neutrons would quickly become enormous and the reaction would become explosive in nature, releasing vast quantities of energy. Such is the case when atomic bombs are detonated.

Setting up chain reactions so that energy may be released in a controlled and safe manner is a great problem for nuclear scientists. To tame the awesome power within these atoms, materials had to be developed that could absorb some of the released neutrons so that they could not cause further fission. In a steadily controlled reaction, each neutron that collides with a nucleus causes the release of additional neutrons. A predetermined number strike other atoms, but most are captured by neutron-absorbing control rods made of cadmium metal.

Another method of controlling the reaction involves regulating the size of the nuclear fuel. It was determined that, if less than a certain amount of fuel were present, the reaction would not be able to sustain itself and would eventually stop. This amount, known as the critical mass, could be formed by bringing two subcritical fuel masses together. To stop the reaction, the masses could be separated.

The first controlled chain reaction was accomplished by Fermi and his co-workers at the University of Chicago in 1942. They utilized uranium as the fuel. The uranium atoms split to form lighter elements and, at the same time, released neutrons that could continue the chain reaction. During the process, some mass was lost and converted to energy. Confirming Einstein's theory, the amount of energy released was quite large.

How Is Nuclear Energy Commercially Obtained?

Nuclear reactors are devices for producing, controlling, and using nuclear energy. The most important material in the reactor is the fuel, which is placed in the center of the reactor, known as the core. Uranium usually serves as the fuel. Natural uranium exists in two forms; however, only one of them is fissionable. One form, having 92 protons and 146 neutrons, is called U-238, 238 being its atomic mass number—the sum of its protons and neutrons. The other form, U-235, has 92 protons and 143 neutrons. (Such forms of elements, having the same number of protons but differing amounts of neutrons, are common in nature and are called isotopes.) U-235 will undergo fission when struck by a neutron; however, U-238 will merely absorb it without splitting. Since natural uranium consists of 99.3 percent U-238 and 0.7 percent U-235, no chain reaction can be sustained unless either modifications of the reactor are made or the percentage of U-235 in the core is increased.

Fermi found that utilizing a material in the reactor to slow down the fast-moving neutrons would prevent neutron capture by U-238 and promote fission of the U-235. This material is called a moderator. Among the moderators in use today are graphite, the metal beryllium, water made with a rare isotope of hydrogen, and natural water. Each moderator has its own advantages and disadvantages. Some factors that must be considered when selecting a moderator are cost of manufacture, performance at high temperature, and durability when exposed to radiation.

The percentage of U-235 in the core can be increased by first separating it from U-238 in natural uranium. This is accomplished using the slight difference in their weights as the basis for their separation. Reactors containing U-235 in pure form, or natural uranium highly enriched with U-235, have no need for a moderator.

The reaction is further regulated by using control rods, as previously mentioned. These are made of neutron-absorbing materials

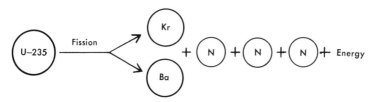

Fission of Uranium-235. An unstable U-235 nucleus splits into Krypton and Barium plus three neutrons and a lot of energy. The total weight of the five products on the right is a little less than that of the U-235 on the left. The missing mass (weight) has been converted to energy.

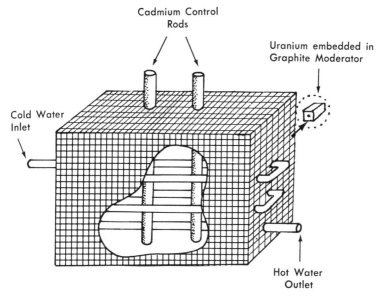

Uranium-graphite reactor pile.

such as the metal cadmium. If the rods are pushed deeper into the reactor, more neutrons are absorbed and the chain reaction must slow down. Pulling the rods out farther allows more collisions between neutrons and U-235 atoms and the faster production of energy.

As the reaction proceeds, the reactor must be kept from overheating. This is accomplished through the use of a coolant. Again, natural water or water made with a rare isotope of hydrogen may be used, as well as liquid sodium and gases. The coolant, flowing through pipes, heats up and is carried away, thus removing much of the generated energy. The heat-containing coolant can be used directly to drive turbines, or it might heat water, converting it to high-pressure steam for generating electricity.

It has recently been proposed that nuclear plants be built in the ocean so that the vast quantities of water can be used for cooling purposes. At the same time, the seawater used for cooling can be distilled into fresh water for use in drought-stricken areas.

How Can the Efficiency of Reactors Be Increased?

The shortage of fissionable materials that can be used as nuclear fuels has concerned scientists for a long time. To help solve the problem, they have improved the performance of nuclear reactors over the years. The most important improvement is the development of devices called breeder reactors. Breeder reactors contain fissionable U-235 together with nonfissionable U-238 that will be bombarded with neutrons, converting the U-238 to plutonium, another fissionable material. The plutonium created in this manner can then be utilized as a fuel itself. A breeder reactor is much more efficient than a conventional reactor, producing more new fuel (Pu-239) than the fuel it consumes (U-235). The equations below describe the breeding and fission reactions:

(1) $U^{238}_{92} + n^1_0 = Np^{239}_{93} + 1$ electron$^-$ (beta particle)
(2) $Np^{239}_{93} = Pu^{239}_{94} + 1$ electron$^-$ (beta particle)

A nuclear-reactor pressure vessel being moved to the containment building (background).

Atomic technicians inspect a nuclear fuel assembly prior to loading in reactor.

Equation (1) describes how U-238 captures the neutron and transmutes to become neptunium (Np). In this process, an electron (beta particle) is emitted. The equation tells us that this electron came from the nucleus of the atom, which is an unusual place for an electron to be. In effect, what has occurred is that a neutron has split to form a positively charged proton and a negatively charged electron. The electron is emitted, while the proton remains in the nucleus, causing the formation of a new element.

Similarly, in equation (2), another neutron in the neptunium atom splits. An electron is emitted, while the proton brings about the transmutation to plutonium.

When struck by a neutron, Pu-239 will undergo fission and release energy. Thus, a fissionable material has been created from a nonfissionable one. A similar method is used to create U-233 (fissionable) from thorium, Th-232 (nonfissionable).

Are There Other Ways of Releasing Nuclear Energy?

The first investigated nuclear reactions were those involving the splitting of heavy elements into lighter ones. During those reactions, there was a loss of mass that had been converted into energy. Subsequently it was found that very light elements may be joined (fused) to form heavier ones, also resulting in a decrease in the total mass. Such a fusion reaction occurs in the hydrogen bomb, which utilizes lithium and isotopes of hydrogen as its fuel. It is also called a thermonuclear reaction.

Hydrogen has three isotopes, all having one proton. Common hydrogen (H_1^1) has no neutrons, whereas the two heavy forms have either one or two. During these fusion reactions, neutrons split lithium atoms to form deuterium (H_1^2) and tritium (H_1^3). These isotopes then combine to form helium (He_2^4) and release tremendous amounts of energy. This is similar to the way in which energy is produced on the sun and other stars.

Although water from the oceans can provide us with a virtually un-limited supply of hydrogen fuel, many severe problems limit our use of this energy source for peaceful purposes. The most significant problem is that an extremely high temperature is needed to start the fusion reaction. In detonating the hydrogen bomb, an atomic bomb serves as the trigger. For peacetime use, this method is impossible.

Scientists have experimented with other ways of producing the 10,000,000° C. temperature needed. In one method, hydrogen is converted to a gas of ionized (electrically charged) particles. Such a gas is known as a plasma. Using electrical or magnetic means, the plasma ions are accelerated. As their speed increases, so does the temperature. The plasma must be controlled so as not to touch, and thereby melt, the walls of its container. One method uses a dough-nut-shaped container called a torus. The torus, surrounded by a magnet, holds a very small quantity of heavy hydrogen gas. Electricity is used to ionize the gas, and a magnet is used to pinch and compress the plasma in the center of the tube. As the compressed ions collide, their

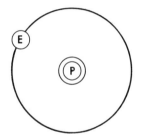

Hydrogen ($_1H^1$) – 1 Proton
1 Electron

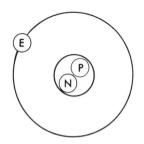

Deuterium ($_1H^2$) – 1 Proton
1 Neutron
1 Electron

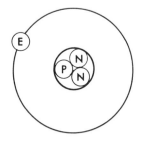

Tritium ($_1H^3$) — 1 Proton
2 Neutrons
1 Electron

Isotopes of Hydrogen.

Controlled fusion research by means of a battery of synchronized lasers uses this vacuum tank to contain the interactions. Pellets of frozen deuterium are dropped into the tank at top while lasers fire their powerful beams through windows at focal point in the center.

temperature increases. Scientists hope eventually to raise the temperature to a point at which fusion will occur.

Another method involves dropping microscopic pellets of frozen deuterium or tritium through the path of high-intensity beams of light known as laser beams. The pellets, sometimes encased in tiny, hollow glass spheres, are frozen in liquid helium. The powerful laser beams first crush the glass, then create enough heat so that fusion can occur.

A great advantage of fusion over fission reactions is that energy can be created with no hazard of radiation. Using either the torus or laser methods, complete success has not yet been reported. An additional problem is that an arrangement has not yet been established for removing the heat generated from the system so that it may be used effectively. Scientists from many nations are working on these problems, and it has recently been reported that the required temperature has almost been reached. However, such a system is not likely to be available before the end of this century.

How Is Nuclear Energy Used?

Although the most prominent use of nuclear energy concerns its involvement in weaponry, its peacetime applications are growing at an ever-increasing rate. A primary reason for this is the diminishing supply of petroleum and natural gas. From an ecological standpoint, many experts believe that the use of atomic energy will eliminate the air-pollution problems caused by the burning of fossil fuels. In addition, scientists have discovered ways of using nuclear energy in medicine, industry, and agriculture.

Approximately fifty nuclear power plants were in operation in the United States in 1974. It is planned to construct about nine hundred more by the end of this century. In these generating plants, each pound of U-235 that is converted to energy will produce the equivalent heat of about 2,500,000 pounds of coal. In other words, a lump of uranium about the size of a golf ball packs as much energy as twenty-five railroad cars full of coal. The energy of nuclear fuels is also being used to provide power for submarines and other ships. In 1958 the submarine *Nautilus* became the first vessel to travel under the North Pole. This was made possible by the fact that nuclear-powered vessels do not need air to burn their fuel, as do conventional ships.

In addition to energy, the reactor has other products. Neutrons from the reactor can be used to bombard atoms of many elements in order to create radioactive isotopes. These radioisotopes are used extensively in medicine. Called tracers, they may be swallowed or injected into the patient's body. With a sensitive instrument known as a Geiger counter, which measures the presence of radiation, doctors can follow the radioisotopes as they travel through the body. Methods of analysis have been developed to diagnose diseases such as cancer in this manner.

In industry, radioisotopes can help inspect the quality of materials, or measure their thickness. Chemists can follow the path and progress of chemical reactions better when a radioisotope is used in place of the nonradioactive element. Similarly, agricultural experts and biologists use tracers to learn more about plant processes and food production. Even the behavior of animals such as insects can be studied by giving them a small dose of radiation and then following their movements with a Geiger counter.

Gamma rays are another product of nuclear fission reactions. These are electromagnetic waves of high frequency. They can be very destructive, presenting a danger wherever reactors are set up. We have, however, been able to put their deadly effects to some good uses. For example, gamma radiation can be aimed at cancerous tissue in the body, destroying it and thereby curing certain forms of the disease. Foods subjected to gamma radiation can be kept for months without refrigeration. The radiation sterilizes the food by destroying the microorganisms that ordinarily feed upon it.

After applying reactor heat to run steam turbines and generate electricity, the coolant temperature may need to be further reduced before it can be recycled. This excess reactor heat can be removed in the distillation of seawater. It can also serve to change coal into much-needed petroleum products and gas fuels. An additional way of using the excess heat is in the direct warming of homes and other buildings in the vicinity of the plant.

What Problems Must Be Overcome?

The most important problem associated with the production of nuclear power is that of radiation hazards. While alpha and beta rays have low penetrating power, gamma radiation can be very harmful, causing burns, death, and mutations (hereditary changes affecting children yet unborn). In order to guard against this hazard, it is necessary to place reactors under shields. Radiation shielding usually consists of materials that slow down and absorb neutrons and materials that are dense enough to stop gamma rays. Water, lead, and concrete are often used for these purposes.

Another major problem concerns the disposal of radioactive wastes. If the waste contains a very low level of radiation, it is diluted in large amounts of air or water and then released. With highly radioactive waste products, the procedure has been to enclose it in shielded containers, which are then buried deep underground. It has been proposed to bury the waste at sea, but that will present a danger to sea life and, eventually, to ourselves. One suggestion has been to launch rockets that would carry the radioactive material into space, but this is also risky because rocket motors may fail and rockets may crash.

Nuclear reactors generate large amounts of heat. The water needed to cool them has been dispersed into rivers, lakes, and oceans. As a result, some harm has befallen fish and other life forms that depend upon moderate and stable water temperatures in order to survive. This problem is known as thermal pollution.

Yet, despite all these drawbacks, the search for ways of using nuclear fuels continues, because the danger of a lack of fuel is still greater to our socioeconomic life than the dangers caused by the use of nuclear fuel.

QUESTIONS TO THINK ABOUT

1. What forces do nuclear reactions release?
2. How did Becquerel find that certain minerals were radioactive?
3. Define transmutation. Why did the discovery of transmutations disturb scientists?
4. What happens to the mass of a radioactive substance?
5. What did Einstein's theory of relativity show?
6. What procedures do scientists use in order to cause the transmutation of various elements?
7. What advantages does the use of neutrons have over the use of protons and alpha particles?
8. What is nuclear fission?
9. What is the cause of a chain reaction?
10. What purpose does a control rod serve?
11. What is meant by a fuel's critical mass?
12. Describe the differences between U-235 and U-238.
13. What is an isotope?
14. What is the purpose of a moderator?
15. Describe how control rods can be adjusted.
16. What purpose does the coolant serve?
17. Write the equations for the production of plutonium from U-238.
18. What is a beta particle?
19. Why do we make Pu-239?
20. What occurs during a fusion reaction?
21. What are the three isotopes of hydrogen?
22. Where do thermonuclear reactions occur in nature?
23. What difficulty must be overcome in order to achieve fusion?
24. What is plasma?
25. What is a laser beam?
26. Why is controlled nuclear energy becoming more important?
27. If one pound of U-235 is converted to energy, how much energy will be produced?

28. What is a tracer?
29. What is a Geiger counter?
30. How can gamma rays be used to help mankind?
31. How can reactor heat be used?
32. What three great hazards must scientists overcome in order to use reactors?
33. What materials are used to stop radiation?

VIII

Energy and the Environment

The environment includes the air we breathe, the water we drink, and the soil upon which we live. In preceding chapters we have discussed how the use of energy has created problems for the environment. Today, many individuals and organized groups are concerned about the harmful effects of excessive or improper energy usage. They realize that the biosphere, those few miles' thickness of atmosphere, soil, oceans, lakes, and rivers inhabited by living things, is very fragile and may be permanently damaged by misuse. These concerned citizens, often referred to as environmentalists, are important figures in the political struggles that are constantly being fought in a democratic country trying to reach decisions about its energy policies.

In response to this justified concern, the Congress of the United States passed the National Environmental Policy Act in 1969 and created the Council on Environmental Quality. One year later Congress passed the Clean Air Act and created the Environmental Protection Agency (EPA) to administer its regulations. In addition, many states and municipal governments created similar agencies for regulating the quality of their areas of jurisdiction.

Many colleges and universities have also demonstrated their awareness of the need for further research and proposals for action in this area by developing interdisciplinary programs under the

101

general title of environmental science. These programs include elements of a variety of sciences, such as biology, chemistry, physics, ecology, meteorology, oceanography, and geology, needed to develop an understanding of the nature of environmental problems.

How Serious Is the Problem of Air Pollution?

Pollution is caused by the addition of harmful impurities to the environment. One estimate of the dollar cost of air pollution is $16,000,000,000 per year, which takes into account deaths, serious illness, agricultural damage, and loss of property values. At a recent conference on "Air Pollution Medical Research" held under the auspices of the American Medical Association, it was reported that children are significantly affected by toxic impurities in the air. As many as 20 percent of the children in polluted environments like those of Los Angeles and New York City may develop acute or chronic lower respiratory disease.

The harmful effect of air pollution on persons already afflicted with respiratory illness has been well documented for a long time. The two groups most severely affected are the elderly and infants. Air pollution can, at times, increase to critically dangerous levels during a "temperature inversion," a stagnant climatic condition caused by a layer of warm air trapped by a layer of cold air above it. The polluting substances caught in the stagnant air can reach fatally high concentrations and produce killer smogs. One of the most infamous examples of such pollution emergencies occurred in Donora, Pennsylvania, killing scores of susceptible people and hospitalizing hundreds over a period of several days. The city of London has also suffered from such severe smog conditions.

Most air pollution results from the burning of fossil fuels. In some cities, incinerators used to burn solid garbage are another source. Where industries are concentrated, the factories, chemical plants, and refineries all contribute to the pall of smoke and soot that overhangs the area. Wherever there are concentrations of automobiles, trucks, and buses, their combined exhausts pour pollutants into the atmosphere. Power generating plants, which must burn thousands of tons of fuel per day in order to produce the electricity we crave, are another source of toxic ingredients. The EPA has established that certain substances are toxic pollutants. These include sulfur dioxide, carbon monoxide, hydrocarbons, nitrogen oxides, peroxides, and particulates.

Since coal and oil always contain varying amounts of sulfur, their combustion results in the release of sulfur dioxide into the air. Sulfur dioxide is a poisonous gas, irritating to the lungs and noxious in odor. Another harmful aspect of sulfur dioxide is that it turns into an acid upon reacting with water, resulting in corrosive rainfalls in industrial areas that deteriorate metals and masonry over a period of years.

Carbon monoxide is the result of the incomplete combustion of carbon because of an insufficiency of oxygen in the burning devices. The internal-combustion engine is the principal source of this highly poisonous gas, which kills by reacting with a person's red blood cells and locking oxygen out of the body. As little as 0.2 percent concentration of this gas in the atmosphere can cause unconsciousness or death. Concentrations approaching this figure are often reached in automobile tunnels and during periods of heavy traffic congestion.

Oxides of nitrogen are an inevitable result of any combustion process. Nitrogen and oxygen are the two major components of air, and the energy released by burning merely accelerates their chemical combination. Many oxides of nitrogen are poisonous, especially nitrogen dioxide.

Particulates are microscopic grains of dust that can irritate the lungs and eyes or trigger allergic reactions in susceptible people. Usually, the particles are composed of carbon as another by-product of incomplete combustion. In this case, the carbon atoms in the molecules of fuel do not oxidize at all. In addition, coal contains varying amounts of minerals that cannot burn. These form a fine dust, called fly ash, which acts together with carbon dust and sulfur dioxide to harm the body's trachea and bronchial tubes.

According to many research studies, the effect of air pollution has been shown to be a factor in causing illness or death. When areas of high pollution (Los Angeles, New York City, Birmingham) are compared with less industrialized areas, it is found that the rates of lung cancer, heart disease, infant mortality, and deaths from bronchitis are higher in the dirty areas.

Today, no one argues whether or not dirty air or water is good for you. Instead, the arguments range over different questions: How clean must the environment be? How much money are we willing to spend? What are the best ways of solving the problems of pollution?

How Can Air Pollution Be Controlled?

Since the main source of air pollution is the burning of fossil fuels, the focus of attack on the problem is directed at the most important consumers of coal, fuel oil, and gasoline. Electric power plants, factories, and other industrial plants are being required by law to meet the standards set by the Clean Air Act and enforced by the EPA. These standards specify how much of each air pollutant may be permitted to escape into the atmosphere each day. If the percentage allowed is exceeded, heavy fines may be levied against the offending company.

Sulfur dioxide, one of the major pollutants, is being counteracted in several ways. First, low-sulfur fuels are being sought and sold at premium prices because of their higher quality. Low-sulfur fuels generally contain less than 1.5 percent sulfur. Natural gas contains little or no sulfur, but it is in too short supply to satisfy our industrial requirements. Much of the crude oils that are low in sulfur content have to be imported at high cost. Crude oil can also be processed to remove the excess sulfur, but this raises the cost of fuel and the energy it provides. Similar methods are being used to remove the sulfur from low-quality coal. The low-sulfur coal in the great unmined areas of the western United States ranging from North Dakota and Montana south to New Mexico is a vast reservoir of energy sufficient to last over 100 years. However, the coal cannot be mined extensively until the environmental impact of this land usage has been sufficiently studied.

Soot, fly ash, and other particulate pollutants can be kept from entering the atmosphere by a process that uses electrostatic precipitators to remove them from the smoke of burning fuels. These devices operate on the principle that oppositely charged bodies attract each other. The main components of the precipitator consist of parallel collector plates attached to the positive pole of a source of high-voltage direct current. As a smoky mixture of dust and gases enters the system, the electrical field gives the dust particles a negative charge. The positively charged collector plates attract the negatively charged dust particles and remove over 99.8 percent of them from the combustion products. As the electrostatic attraction continues, a heavy layer of dust collects on the plates. Periodically, the plates are vibrated and the particulate matter falls off into a collection hopper and is carted off. The remaining gases rise into the air through the smokestack, but the emissions are practically invisible because of the absence of soot and fly ash.

The EPA wants large factories and electric-generating plants to install special devices, called scrubbers, to remove most of the sulfur dioxide from their chimney gases. The scrubber uses a liquid spray to remove the sulfur dioxide by absorption or chemical reaction. The main principle of the scrubber is to use a basic compound, such as lime or magnesia, to react with and neutralize the acidic sulfur dioxide. Scrubbers are still being tested and evaluated to see if their remaining technical difficulties can be overcome, before requiring industries to invest hundreds of millions

Combination photograph and cutaway drawing showing how dust particles collect, through electromagnetism, on charged plates inside an electrostatic precipitator.

of dollars installing them. Fortunately, preliminary results appear promising, and scrubbers remain a possibly potent weapon against one of the worst sources of air pollution.

Automobiles and trucks were first required in 1973 to meet

EPA standards for the emission of air pollutants. For 1975-model cars, the limits that could be emitted per mile were 1.5 grams of hydrocarbons, 15 grams of carbon monoxide, and 3.1 grams of nitrogen oxides. The standards set for California, where air pollution is very severe, were 40 percent stricter. In 1974 all new cars sold in the United States were equipped with various pollution-control devices to meet the new standards. The catalytic converter was one of the most widely used methods for changing pollutants into harmless by-products. One result of its use was the requirement of a new lead-free gasoline in all service stations to enable the converter to function effectively. Lead compounds in the gasoline tend to foul the catalyst and clog the converter.

What Is Thermal Pollution?

Thermal pollution is the warming up of the atmosphere or natural bodies of water by waste heat coming from large factories or power-generating plants. Problems arise because many industrial concerns and almost all electric utilities deliberately choose a factory or power-plant site near water in order to use it for cooling purposes. Cool water drawn from the neighboring stream, river, or lake is sent to a condenser, where it absorbs the waste heat of the power production equipment. The cooling water becomes much warmer and is then recirculated back into the natural source from which it was originally drawn. In many cases, lakes or rivers are too small to reabsorb the warm water without a significant increase in normal temperature.

Environmentalists believe that raising the temperature of lakes, rivers, or streams even a few degrees can have harmful effects on the spawning process of the fish population. In addition, an increase in the water temperature can cause algae to grow so thickly in the water that some species of fish are unable to live. A third serious possibility is that a combination of higher water temperature and increased algae concentrations can cause a lowering of the oxygen content in the water, with the result that many fish die of suffocation.

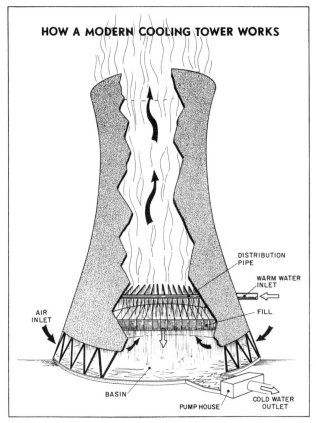

A diagram of a cooling tower, showing its principles of operation.

The hyperbolic-shaped cooling tower has been adapted to solve the most serious aspects of thermal pollution. Originally developed in Europe, where lakes and rivers were too small to supply the demand for large amounts of cooling water by power plants, it is used in the United States to send the waste heat absorbed by the cooling water into the atmosphere rather than into the natural bodies of water.

COURTESY AMERICAN ELECTRIC POWER CO.

The cooling towers at the General James M. Gavin power plant at Cheshire, Ohio.

The principle by which the towers operate is the cooling effect of evaporation. Warm water from the condensers is piped into the tower and flows down over a series of baffles. These break up the water flow and promote evaporation, a process that cools the remaining water. The cooled water then collects in the basin of the tower, where pumps can recirculate it back to the condenser or return it to the lake or river source without causing thermal pollution.

The purpose of the hyperbolic shape of the tower is to get rid of the heat that escapes during the evaporation process. Warm, moist air forming around the baffles expands. Cooler, drier air moves into the tower through its open bottom, creating an updraft. The special shape of the tower is designed to focus the updraft into a rising column of humid air that will penetrate high into the atmosphere in order to minimize alteration of the meteorological conditions in the vicinity of the tower.

Some environmentalists have predicted that cooling towers would create as many new problems as they were designed to solve. It was thought the towers might produce local ground fog, increase the humidity, or cause rain or snow storms in the area. Another fear was that icing of highways, bridges, or power lines might occur in frigid weather. Fortunately, none of these situations has happened. The American Electric Power Company, the largest private electric utility in the United States, reported on a ten-month study and series of tests carried out on the cooling towers at their generating plants. The only effect observed was a 1 percent increase in the relative humidity downwind from the towers. The company concluded that cooling towers are effective devices for controlling thermal pollution without creating new difficulties of any significant degree.

Is Nuclear Power Safe?

Following World War II, photographs showing the horrifying effects of radiation on the victims of the atomic bombings of Hiroshima and Nagasaki were published in newspapers and magazines throughout the world. Many books, articles, and research studies dealing with the subject have also been widely circulated. By now, most people are aware of the dangers of radioactivity.

Radiation from nuclear fuels or their waste products consists of alpha, beta, and gamma rays. Alpha rays (helium nuclei) and beta rays (electrons) are not very penetrating particles: the latter is unable to travel more than a few feet through air; the former, barely one inch. Gamma rays are pure radiant energy of very short wavelengths and are extremely powerful and dangerous. It takes a sheet of lead several inches thick or a wall of concrete over three feet thick to stop gamma rays.

Radiation is dangerous to the body in two ways. One way is directly damaging to the person by the immediate effect of the penetrating rays, which destroy the complex enzyme molecules in the cells of the body. Although one particle or ray will probably damage only one cell, should the dosage of radiation be great, a large enough number of cells can be affected to bring on radiation sickness. Very severe doses result in death. Sometimes, affected cells may not show symptoms immediately, but later they may develop cancer or related illnesses. Also, the accumulation of small amounts of radiation can eventually add up to a heavy enough dosage to cause illness at a later time.

The second effect of radiation on the cells is to alter the fragile chemical nature of the chromosomes, the carriers of the human genetic heritage deoxyribonucleic acid (DNA), located in the nucleus of each cell. These slender double helical strands of DNA molecules contain the chemical blueprints of bodily structure and function, and are duplicated and reduplicated from generation to generation. When any molecular change occurs in the DNA, a mutation may appear in the next offspring, which develops according to the altered hereditary blueprint. Almost always these mutations are harmful, resulting in physical deformities or inabilities to carry out normal life functions. The rate at which mutations are produced is in proportion to the dosage of radiation exposure. In other words, unnecessary radiation exposure is to be avoided.

As a result of these effects, environmentalists are deeply concerned about the safety of nuclear-powered electric-generating plants. Two related fears are the danger of a nuclear explosion or the accidental release of radioactive materials. Another problem involves the disposal of the radioactive waste materials produced by spent fuel elements.

Fortunately, there is no chance of the core of a nuclear reactor exploding like an atomic bomb. The design of the device makes it impossible for a critical mass of fuel to concentrate itself in a short enough time to detonate. However, the fuel could theoretically overheat sufficiently to fuse the core and containing material into a white-hot mass capable of melting its way into the ground to a considerable depth. This could breach the containing shell and allow radioactive materials to reach the atmosphere, where the dangerous fallout could be carried up to 100 miles downwind, threatening thousands of people with radiation sickness or death.

The fear of such an accident has led many people to oppose the building of a nuclear-powered generating plant near them. By organizing themselves into effective pressure groups, such people have postponed or blocked the plans for construction of many such power plants.

The accidental release of radiation caused by overheating of the core is normally prevented by the flow of coolant material that transfers the heat to the electric-generating units. The most likely causes of overheating are the failure of the control rods to slow down the chain reaction, or the loss of coolant needed to carry off the excess waste heat because of a burst pipe. If either fails separately, the other can be used to moderate the heat and prevent both serious damage to the structure and any release of radiation. In addition, back-up cooling systems are always standing by to cool the core in an emergency. Experts at the Atomic Energy Commission place the odds against the breakdown of all the protective systems simultaneously at one million to one.

In fact, since nuclear generators began operating in the 1950's, accidents have caused seven deaths to workers, and none to consumers. One fuel melt-down did occur, but it was successfully contained by the concrete shell so that no radiation escaped. By contrast, in the United States, over 40,000 people are killed annually in automobile accidents, 200 or more die in commercial airplane accidents, and thousands are made ill or die from air pollution caused by the burning of fossil fuels. In comparison, nuclear-energy projects have enviable records of safety. The Nobel Prize-winning physicist Dr. Hans Bethe, who pioneered in the study of fusion reactions, believes the benefits of atomic-power plants far outweigh the possible dangers involved in their use.

The disposal of spent fuel elements and other radioactive waste materials produced by a nuclear reactor without causing environmental pollution is another headache for nuclear-energy enthusiasts. The major problem is to make sure that radioactive substances remain sealed up for a long enough time to "cool off" by having their radiation rate slow down to a safe level. The other problem is to find economical ways of accomplishing this for ever-increasing amounts of "hot garbage" as we put more and more nuclear power stations into operation.

Some of the less dangerous wastes have been dropped in the sea sealed in concrete vaults designed to last 100 years. The most radioactive garbage is sealed in steel and concrete vaults and buried deep underground. Liquid wastes are a particular problem because they might corrode their containers, leak out, and contaminate groundwater supply. One experimental method that has been designed is to use clay to absorb the "hot liquid waste," then bake the impregnated clay to seal the radioactive liquids within it. The clay can then be safely buried.

A novel plan has been devised for handling the greatly increased amount of radioactive wastes that will be produced in the future. These wastes will be buried in abandoned salt mines deep in the earth, where they can be monitored for leakage at regular intervals. Another plan involves pumping the waste into leakproof caverns created by nuclear explosions deep in the earth. The head of the Atomic Energy Commission contends that this technology will be safe for storing the approximately 135 metric tons of radioactive waste expected annually by 1985.

Such statements are of no comfort to dedicated environmentalists and consumer advocates. Ralph Nader, perhaps the best known of this group, has argued that nuclear-power plants should be stopped dead in their tracks until reactors are safe beyond doubt. Nader argues that if the choice were between nuclear power or candles, the people would choose candles if they knew the facts. The issue remains unresolved and becomes a matter for the people to decide, based upon the best information they can get and with a clear picture of the available alternatives.

COURTESY ATOMIC INDUSTRIAL FORUM, INC.

An underground room in a salt deposit 1,000 feet deep is being tested for the safe storage of radioactive wastes in stainless steel canisters.

Can We Stop Defacing the Land?

The Appalachian mountains rise in Tennessee and run north through Kentucky and West Virginia to Pennsylvania. Nine states in all comprise this hilly region known simply as Appalachia. Large parts of it are coal country, and tunnels bore through the seams of bituminous coal or anthracite to extract the valuable mineral. In many areas where the coal lay close to the surface, strip-mining techniques were used to dig out the coal at the lowest possible cost. Once the mineral deposits were exhausted, the mine owners abandoned the pits and left the land in a ravaged and useless condition. In 1974, there were 2,000,000 acres of ugly, barren "orphan spoil banks" left as a remainder of the strip mines and the greed of their operators.

Environmentalists cite the evidence of this destruction of land as proof that strip mining should be banned unless it is strictly regulated by state and federal legislation. In response to this demand, several states have passed laws requiring strip-mine operators to post cash bonds and plans for reclaiming the land and restoring it as closely as possible to its original condition. In addition, the Congress has approved legislation on a national level setting forth similar requirements. However, large elements of the coal-mining and electric-utility industries oppose such legislation. They argue that overly strict controls will delay the vitally needed increase in coal production required to overcome the nation's energy shortage. These groups have the power, financial strength, and organizational ability to form lobbies to influence legislators to pass bills favorable to their interests in minimum controls. The past history of Appalachia is testimony to the power of such mining interests. But times change, and environmentalists are also organized, financed, and powerful. The struggle between the opposing opinions is no longer one-sided.

Fortunately, it is often possible to extract coal economically by strip-mining techniques and then restore the land to a useful condition. This has been demonstrated on over 300,000 acres of reclaimed land as of 1972. As an example, the Ohio Department of Natural Resources requires land-restoration plans and schedules before mining begins. The plans must show what the land is like at the start, how it will be graded and contoured, how lakes and ponds will be created from the pits, and how it will be reforested and vegetated after restoration. It is only when the plan is approved, and bonds as high as $2,800 per acre are posted that a mining license is issued and actual operations can begin.

The benefits of restored lands are enormous. Erosion of the spoil

COURTESY AMERICAN ELECTRIC POWER CO.

Strip mining for coal devastates the land.

COURTESY AMERICAN ELECTRIC POWER CO.

The land ruined by strip mining can be restored to its original beauty and value.

banks is halted, and in place of naked hills are pastures and rangelands suitable for grazing. Reforestation produces woods that are wildlife habitats as well as a renewable lumber resource, in place of a nonrenewable mineral resource. Recreation lands and parks with ponds and lakes suitable for camping are often created. In some cases, the restored land can be cultivated if the soil and climate permit.

Environmentalists are also concerned about energy developments in Alaska, our last great frontier of natural wilderness. The discovery of oil in 1968 on the North Slope near Prudhoe Bay was welcome news to an energy-hungry nation, but it also meant that a fragile frozen land was endangered by the need for industrial exploitation. The halls and committee rooms of the Congress echoed to the arguments for and against the development of the oil. Environmentalists showed how delicate were the mosses, lichens, sedges, and other dwarf vegetation of the arctic area. Ecologists demonstrated how the long winters and the short summers could turn any destruction of the land into scars that could take decades or centuries to heal. But the energy crisis was too great, and the decision was made to proceed in the development of a huge new oil field while minimizing the environmental impact as much as possible.

The next problem was to transport the petroleum to its markets. Would ships or pipelines do the job best? Ice-breaking tests were carried out using the tanker *Manhattan,* refitted with a reinforced bow and heavily instrumented to measure the stresses and strains encountered on its trip through the icy Northwest Passage to Alaska. Environmental-impact studies were made of various pipeline routes and their effects on the tundra, permafrost, and wildlife of the region. At last, after four years, the decision was made. It would be a pipeline running 800 miles across Alaska from Prudhoe Bay on the North Slope, across the mighty Yukon River, past Fairbanks, to Valdez, a port on the ice-free Gulf of Alaska.

The potential impact of the trans-Alaska pipeline on the envi-

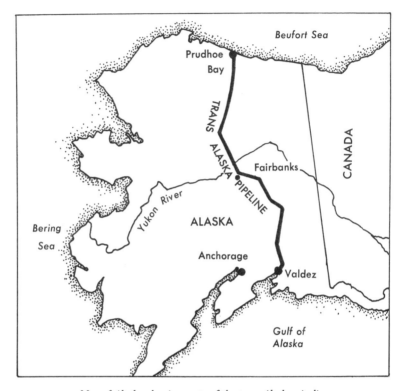

Map of Alaska showing route of the trans-Alaska pipeline.

ronment is great. One effect results from the construction work along the pipeline route. Heavy trucks and bulldozers cutting through the thin layers of arctic plants to the soil below produce scars in the land that are slow to heal with new growth and quick to erode in the short summer thaw. Botanists working for the pipeline company have sought rapidly growing species of grasses that would re-cover the bare earth with soil-holding vegetation. Fortunately, twenty such species have been reported found.

Another source of potential trouble is the temperature of the oil, which must be kept at 145° F. to keep it flowing. If the heat reached the permafrost, or solidly frozen soil lying a short distance under the surface, it would thaw and turn into a soupy mud, unable to support the pipeline. Should one of the pipeline joints crack open, thousands of barrels of crude oil would spill over the landscape, fouling plants and animals and polluting soil and streams. This is to be prevented by building a raised gravel path, up to five feet thick, for the pipeline to rest upon and to insulate the permafrost from the hot oil.

Alaska has many earthquakes because it rests on top of an unstable portion of the earth's crust. A severe tremor could possibly rupture the line or the oil-storage facilities at its terminus in Valdez. Designers have planned the structures to withstand shocks of up to 8.5 on the Richter scale without suffering severe failure. Extra valves have been installed to minimize any oil spill if the unexpected should occur. Earthen dikes built around the storage tanks will retain any leakage.

The unique animal life of Alaska and the majestic beauty of its frozen wilderness rising from the rolling tundra to the highest mountains in North America were also threatened by the pipeline and its related construction. Environmentalists raised many questions. Would the migrations of the caribou herds as they grazed for food be blocked by the raised ledge supporting the pipeline? Would the highway paralleling the pipeline route open up the wilderness to the worst of human exploitation, such as highway litter, hunting, and tearing up the tundra with dune buggies or similar equipment?

There is no way to guarantee that some harmful practices will not occur in Alaska and result in some hardship to the wilderness. But it is also true that if the pipeline were not built, there would be very real economic hardships for the people of the United States, who would suffer from industrial slowdowns caused by energy shortages, unemployment, and lowered incomes and standards of living. Once oil begins to flow through the pipeline, Alaska is expected to collect royalties and taxes from it amounting to more than $1,000,000,000 per year by the 1980's. The Native Claims Act of 1971 gives Alaska's Indians and Eskimos the right to reserve 40,000,000 acres of land for their traditional way of life. It also starts the process of setting aside 32,000,000 acres of new national park land in the state. As William Egan, governor of Alaska said, "I think the pipeline is a model for the state and the nation, because it shows we can develop our resources in a safe and acceptable way, environmentally."

As in the case of the trans-Alaska pipeline, other proposals for building new facilities for producing and transmitting energy generally arouse considerable opposition from local environmentalists. Pipelines, generating-plant site locations, pumped storage projects, new hydroelectric dams, and additional transmission-line routes suffer long delays from legal suits and demands for additional environmental-impact studies, all emanating from the environmentalists' concern for the preservation of natural resources.

Electric-power transmission lines are an example of how continuing research and development can help us use energy efficiently and conserve the environment as well. The rights-of-way for such lines occupy considerable public and private land, the wide swaths of their cross-country routes running over hills and valleys and through forests. Since 1916, high voltages for bulk power transmissions have risen from the original 138,000 volts to 345,000 volts in many areas. More power can be carried over the higher-voltage lines. The American Electric Power Company has developed an extra-high voltage line (765,000 volts) for distributing power through its seven-state system. The same company also has plans ready for transmission lines of ultra-high voltages (1,500,000 volts). One such line would be sufficient to carry all the power needed to illuminate Boston, Pittsburgh, St. Louis, Baltimore, and Washington, D.C. As a result, it could replace up to 150 rights-of-way used for 138,000-volt lines. The use of ultra-high voltage lines will need only one-fifteenth of the land currently required for transmission line rights-of-way, a significant conservation of the wilderness.

In this book, we have been able to discuss only some of the ways that our use of energy has defaced the land. The many other examples of harmful environmental impact can be read about in other sources. Fortunately, the land's ability to recover from the damage inflicted by thoughtless deeds is great. As we have indicated, the work of reclamation has begun, and its successes can already be measured in significant degree. We can, with proper planning, improve our energy reserves and protect and maintain our environment.

Can We Make Our Water Pure Again?

Many factors contribute to water pollution. It is not only an industrial problem, but also a people problem, caused in large part by the increased use of chemicals in our daily lives. Soaps and detergents containing phosphates end up in the drain water and flow through sewers to join our streams, lakes, rivers, or oceans. These nutrients cause algae to multiply rapidly. The farmer sprays fertilizers and insecticides on his fields and plants, and the rains eventually wash them into the rivers. Thousands of fish may die as a result of this poisoning of the water.

Many towns and cities permitted their raw, untreated sewage to flow into the local stream or river, turning them into open cesspools. Also, many industrial plants chose sites near a source of water to meet their technical needs for cooling or cleaning purposes. They would then discharge their used, polluted water back into the source without any thought of its environmental effects. To many industrialists, the water, like the air, was free.

The oceans, being the largest bodies of water on the planet, were the last to be thought of as susceptible to pollution. For decades, New York City sent bargeloads of trash and treated sewage 20 or more miles out to sea to be dumped, only to discover that even the ocean was not big enough to absorb so much waste. A creeping tide of thick black gooey muck has spread across the more than 20 miles of ocean floor to endanger the miles of beaches on the Long Island shore.

A major source of ocean pollution is oil spills. The millions of barrels of crude oil transported daily from the oil-producing nations to the oil-consuming nations travel by ship. Today, these are the largest ships that man has ever built, supertankers that make the huge passenger liners of yesteryear seem small by comparison. The largest of these tankers carry 280,000 tons of crude oil. Even larger ships capable of hauling 400,000 tons of crude oil are on order.

COURTESY EXXON CORPORATION

Supertankers, the largest ships afloat, carry up to 400,000 tons of crude oil.

Such huge vessels must ride low in the water for maximum stability. There is no problem when they are filled with oil, but after they discharge their cargo, they ride high out of the water. To compensate, they fill their empty cargo tanks with seawater, which acts as a ballast, and settle down to their normal displacement depth.

Problems arise when they must pump out the water to refill

their tanks with fresh cargo. A fairly large amount of oily residue comes out with the water and forms a polluting oil slick on the ocean. At other times, it becomes necessary to clean a cargo tank by flushing it out with water, again discharging the oily wastes into the sea.

The most catastrophic oil spills occur when a tanker is involved in a collision with another ship, grounds itself on a reef breaching the hull, or breaks up in heavy seas during a storm. In one famous case the *Torrey Canyon,* a large tanker, broke up on the rocks off the English coast, creating a massive spill that formed oil slicks many miles long, coating miles of English and French beaches, and killing thousands of seabirds.

Another dramatic example of the environmental risks associated with increased energy use occurred in the Santa Barbara channel along the coast of California, at an offshore drilling rig. The drill struck oil, but it brought gloom, not glee, to the area. Instead of the oil flowing through the intricate maze of pipes meant to direct it into storage facilities, it oozed out of the sea bottom through a large crack in the underlying rock strata. Huge masses of crude oil floated to the surface and washed ashore on the beautiful beaches, fouling the sand and killing seabirds and other sea life in the intertidal zones. The leak was finally plugged with a special hydraulic cement, but not before much damage was done.

Fortunately, through the efforts of many aroused environmentalist groups, the public was made aware of the serious problem of water pollution, and governmental action on both the federal and state levels occurred. Pollution-control laws were passed requiring the treatment of sewage and other industrial effluents. Many industrial plants have since built water-treatment systems to purify their waste water before they discharge it back into the local stream or river.

As a result of these efforts, the tide has turned in the fight against water pollution. In rivers that were once little better than open sewers, it is now possible to catch fish again. The Hudson River in New York City contained only a hardy species of eel in 1968, but by 1974 local fishermen were beginning to catch shad, bass, and other related fish species that had not been seen in the river for 50 years. In England, salmon have been seen in the Thames River. The waters of the world are being cleaned, but the job is far from complete.

QUESTIONS TO THINK ABOUT

1. What is meant by the environment?
2. How would you define the biosphere?
3. What is the major concern of an environmentalist?
4. Describe two actions taken by governments to aid the environment?
5. How can you prepare for a career in environmental work?
6. What are two serious effects of air pollution?
7. Under what conditions does air pollution reach crisis proportions?
8. What is the most serious cause of air pollution?
9. Name three substances that are toxic air pollutants.
10. Why is sulfur dioxide able to corrode metal?
11. Why should you always have good ventilation when burning fossil fuels?
12. What are particulates?
13. Describe two methods of controlling air pollution.
14. In what way is a phonograph record similar to an electrostatic precipitator?
15. What is the purpose of a scrubber? By what principle does it work?
16. How are automobiles kept from contributing to air pollution?
17. What is meant by thermal pollution? How can it cause harm?
18. Describe the function of a cooling tower. Explain how it works.
19. Why does a cooling tower have a hyperbolic shape?
20. What are two ways in which radiation is dangerous to the body?
21. How could a nuclear reactor endanger the environment?
22. What safeguards have been built into nuclear reactors to prevent accidental release of radiation?
23. Describe two methods for safely disposing of radioactive waste materials.

24. Describe the steps some state governments have taken to make strip mining less destructive of the land.
25. What are some of the benefits we gain by enforcing the restoration of strip-mined land?
26. Explain why Alaska is especially vulnerable to environmental damage from oil field and pipeline development.
27. Describe two precautions being taken by the pipeline builders to prevent oil spills.
28. Explain how pollution can often be caused by ordinary citizens rather than by industry.
29. What would be the economic consequences of not building the trans-Alaska pipeline?
30. How does increasing the voltage of electric transmission lines help save the use of land?
31. What are three sources of water pollution?
32. How do oil spills occur?
33. What can be done to stop water pollution?

IX

The Energy Crisis

The requirement of having large amounts of energy available in order to sustain or improve an acceptable standard of living has led to an energy crisis in many countries, including the United States. An energy crisis consists of a shortage of energy severe enough to disrupt the normal functioning of society. Two factors that have combined to produce the crisis are the ever-increasing demand for energy coupled with the present inability to find sufficient new sources of energy to meet that demand.

The signs of the crisis in energy are numerous. Shortages of electricity in many areas of the U.S. require the power utilities to lower voltages during periods of peak demand, creating "brownouts." In some areas, the shortage of electricity has been so great that varying sections of the power grid have been cut off completely from the electrical supply, producing local "blackouts." Homeowners are warned of shortages to be expected in the supply of fuel oil during the winter heating season. Natural-gas companies refuse to accept new accounts because they have only enough supplies to satisfy established customers. More gasoline stations are closed on Sundays and at night because they can sell all of their monthly gasoline supply during normal business hours. You may have seen long lines of automobiles form at the gasoline pumps during times of shortage. Many service stations have gone out of business altogether.

The economic consequences of the energy crisis have had a severe impact on prices the consumer must pay for fuel. Between 1970 and 1974, retail prices of gasoline have doubled. In many areas, the electric and gas bills of homeowners and apartment dwellers also doubled and in some cases tripled. The price of a ton of coal rose from $30 to $50. Imported crude oil jumped in

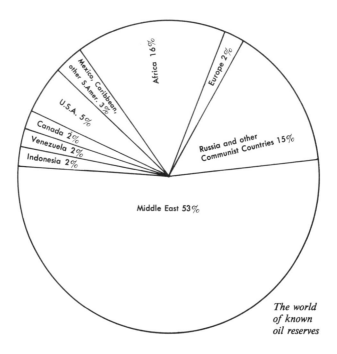

The world of known oil reserves

cost from $3 per barrel to $14 per barrel. All these price rises resulted from the realization that energy demand was increasing faster than the discovery of new fuel supplies to meet future needs.

The U.S., which had been self-sufficient in petroleum supplies for decades, began to require increasingly large imports of crude oil and natural gas from the Middle East, Africa, Venezuela, and Canada. When the oil-exporting countries realized the growing power of their position, they formed a cartel, or international monopoly, called the Organization of Petroleum Exporting Countries (OPEC). By acting as a unified group that controlled 55 percent of the world's petroleum reserves, the OPEC countries were able to manipulate the supply and price of crude oil needed by most of the non-Communist countries.

COURTESY EXXON CORPORATION

Kharg Island, Iran. Iran is a member of the Organization of Petroleum Exporting Countries (OPEC). Tankers carry its oil to the consuming nations.

Since the most important OPEC countries, those with the largest reserves of petroleum, were located in the Middle East, they used their power to influence the politics of that region. This was especially true in regard to the conflict between Israel and the Arab countries. In 1973, after the outbreak of the October "Yom Kippur" war between Egypt, Syria, and Israel, the Arab members of the OPEC group embargoed oil shipments to those countries aiding, or merely expressing support for, Israel. In particular, the U.S. and the Netherlands were severely affected. Other countries such as Japan, Great Britain, France, Italy, and West Germany also suffered from a decrease in shipments of petroleum, but to a lesser degree because in varying ways each of these countries yielded to the pressure exerted by the threat of an Arab oil embargo against them.

Once an artificial shortage of oil had been created by the Arab oil embargo without any show of resistance by the majority of the oil-consuming nations, the OPEC group was able to raise the price of a barrel of crude oil by 400 percent. This meant that the U.S., which imported 1,200,000,000 barrels of crude oil in 1973, would have to pay an oil bill to the OPEC countries for 1974 of close to $16,800,000,000 compared to only $5,000,000,000 in 1973. Other oil-consuming countries faced similar huge increases in their oil import bills; but unlike the U.S., most of these countries were ill prepared to pay them and faced international bankruptcy unless monetary aid was forthcoming.

As a result, in 1974 the citizens of the U.S., in addition to people in other Free World countries, suffered through the worst year of inflation in their history as prices rose an average of 11 percent. The increase in the price of oil and all the goods affected by that increase has been estimated to account for one-half of the inflationary rise. Other countries were even harder hit by the price increase in oil. Worst off were the underdeveloped nations whose plans for improving their people's standard of living were blocked by the expensive oil bills they now had to pay.

To put the problem into more understandable terms, every day in the U.S., the average person uses 4 gallons of oil, 300 cubic feet of natural gas, 15 pounds of coal, and smaller amounts of other forms of energy. The average person in the rest of the world uses only one-eighth this amount but would like to use as much in order to raise his standard of living to our level. World petroleum demand has been rising at an annual rate of 7 percent, and the total amount of oil needed to satisfy this appetite would double by 1985. Since oil and natural gas accounted for 65 percent of total world energy consumption in 1973, the ability of the OPEC nations to exert economic and political influence and power remains formidable. This situation will remain a fact until at least 1985, because it will take that long at a minimum to develop sufficient alternate sources of energy such as coal and nuclear power.

COURTESY EXXON CORPORATION

An oil field on Lake Maracaibo, Venezuela. Venezuela exports much of its oil to the United States; it is also a member of OPEC.

The energy crisis affects almost every sector of life in the oil-importing countries. In the poorest countries, the shortage of energy available for economic-development projects will condemn the masses of the population to continue a subsistence existence. Although modern medical techniques may reduce death and illness, high birthrates and a growing population will mean that hunger and possible famine will be the rule rather than the exception.

In the technologically developed countries, higher prices for energy will raise the price of everything produced by energy. Many luxuries that have been taken for granted may be eliminated. Some people may find the cost of operating an automobile too expensive to continue. Gasoline and other fuels may be rationed or heavily taxed by governmental order to discourage their use. Industries whose business is the production of nonessential items may be denied fuel and other raw materials needed for more vital goods and services, and as a result may have to close down. The number of unemployed people may increase. Food prices will certainly go up, because farmers are heavily dependent on fuel for their tractors and other farm equipment. People in the U.S. will have to spend an ever-increasing percentage of their income on this necessity. In general, an energy crisis will cause a decline in everybody's standard of living, but this would have its severest effect on the people with the lowest incomes.

The energy crisis has rearranged the wealth of the world. The total money paid to the OPEC nations in 1974 by the oil-consuming countries was between $55,000,000,000 and $60,000,000,000. The same amount is predicted for 1975, with somewhat lesser amounts thereafter, assuming that the price of oil is not arbitrarily raised again. Arthur F. Burns, chairman of the Federal Reserve Board, warned that this massive redistribution of economic and political power carried dangers for the future of the U.S.

It is conceivable that a prolonged energy crisis that is not resolved by a spirit of cooperation and compromise between the energy-importing and energy-exporting nations could result in international conflict and war. Such a catastrophe might occur if the people in the developed energy-consuming countries remain ignorant of the nature of energy and its use, and demand from their governments a supply of energy equal to their past rate of consumption. Understanding the limitations of energy supplies and the necessity for developing new habits so that we may learn to live within the energy supply available could help prevent such an awful result.

What Are Some Possible Solutions to the Energy Crisis?

No easy answers have yet been found to solve the energy crisis. In fact, different people have proposed totally opposite ideas for adjusting to the situation. One viewpoint sympathetic to the OPEC countries holds that oil prices are high because the U.S. has supported Israel in its conflict with its Arab neighbors, and that if the U.S. would withdraw its support from Israel, the Arabs would supply us with all the oil we need at reduced prices. Most people feel, however, that it would be suicidally dangerous for the U.S. to allow itself to become dependent on sources of oil that have already proven to be unreliable.

Extreme environmentalists represent another point of view. They disapprove of any action that develops a natural resource while risking any damage to the biosphere, no matter what the social and economic cost of not developing the natural resource might be. Their answer to the energy crisis is for all of us to return to a simpler life and live close to the land as most Americans did in the early 19th century. However, there are more than 200,-000,000 Americans today, and the transition for the majority of them would be impossible without severe privation and suffering.

Another extreme point of view holds that since we are so militarily powerful, the U.S. should occupy the oil fields, which were developed with our money and technical expertise even though located on other nations' territory. Most people believe this would be an immoral act unworthy of a great nation.

Perhaps a balanced, more moderate approach will be most successful in finding an answer to this problem. One way to alleviate the energy crisis might be to find and develop new sources of energy to relieve the U.S. of its dependence on imported oil. To

ــimulate this, the high price of fuels would have to be maintained as an incentive to pay for the expensive risks in prospecting. Scientists and engineers are already seeking new deposits of our present fuel sources in ever more isolated areas. The search for oil and gas now extends out to sea in the waters of the continental shelves. Coal deposits in the Western prairie states, once considered too distant from the centers of population, are now being mapped, surveyed, and leased in preparation for mining. The building of nuclear-powered electrical generating plants may occur at a more rapid rate.

Hydroelectric dams may be built in the few remaining regions where they are feasible. Tidal power can be harnessed in a few suitably located places. Oil shale and tar sands can be mined to provide additional sources of petroleum. Windmills might be used to generate electricity in some areas. In certain volcanic localities, geothermal power plants can be built. Solar power seems attractive to many people, although its practicality appears limited. However, it is not essential that any one approach to solving the energy crisis be the final answer. If each source is able to contribute a reasonable supply of energy, the cumulative result might be sufficient to close the energy gap between domestic demand and supply, making the U.S. self-sufficient in energy once again.

Probably most informed people would agree that the ultimate answer to the energy crisis lies in the development of controlled thermonuclear or fusion power. This would require a scientific and technological breakthrough that is yet to be achieved, but if it were accomplished, we would be able to obtain all the energy normally released by a hydrogen bomb. Instead of a destructive explosion, however, the almost unlimited energy would be released gradually and used to generate electricity sufficient to meet all our needs.

Another way to make the energy crisis less severe is through conservation. President Gerald Ford asked Americans to reduce their imports of foreign oil by 1,000,000 barrels per day by 1975 through voluntary conservation efforts. At the same time, he directed the automobile manufacturers to develop new engines with a 40 percent improvement in gas mileage by 1979. Federal Reserve Board Chairman Arthur F. Burns strongly suggested before a Congressional committee hearing on the energy crisis that conservation efforts of austerity proportions were needed. He indicated that it would be necessary to have less pleasure driving, even if it meant lost business for the recreation industry and reductions in the sales of automobiles.

It seems likely that we must reexamine every use of energy in order to determine how essential it is and to establish a system of priorities. At the same time, energy for essential purposes must be used as efficiently as possible. For example, the vital uses of energy for transportation cannot be denied, but our current use of energy for moving people and goods is most wasteful. Mass-transit systems are more efficient than individual automobiles for moving large numbers of people from one place to another. Similarly, trains are less wasteful of fuel than trucks in moving freight over long distances.

Developing solutions to the energy crisis may require changing our concept of economic "growth." To a large extent, the American economy has been based on the assumption that it is best to design an ever-increasing supply of new items that rapidly become obsolete and are disposed of in refuse dumps. We need to examine the effects of expansion of material goods, suburban housing, and population. Perhaps by such studies, we may gain an understanding of the restrictions on economic growth caused by our limited supply of raw materials.

The purpose of reducing growth is to lower the demand for fuels and energy, thereby conserving the available supply until new sources of energy are developed. Perhaps the answer to the crisis will be found in the gradual reduction of energy demand by limiting population growth, lowering goals for an ever-rising standard of living, and achieving a state of equilibrium between human beings and the environment.

One requirement that can never be forgotten is the need to satisfy our energy needs without damaging the environment permanently. President Ford asked the Congress to amend the Clean Air Act in order to permit lower-grade coal with a higher sulfur content to be burned when weather conditions permit. The President also wants the passage of surface-mining legislation that will balance environmental protection with the development of an adequate fuel supply. Although strict environmentalists might object, the essential fact is that the nation's basic commitment to a cleaner environment is being met. The timetable for achieving that goal is being extended to ensure the economic and strategic security of the U.S. by maintaining sufficient energy availability.

Solving the energy crisis will be the great challenge for the U.S. in the years and, possibly, decades to come. It is the government's role to set national goals and objectives, formulate policies, and reconcile conflicting interests. But ours is a democratic country, and its strength rests on an informed citizenry. This book should not be an end for you, but a beginning to stimulate your interest in the greatest problem of the final quarter of the 20th century.

QUESTIONS TO THINK ABOUT

1. Describe what is meant by an energy crisis.
2. Explain what causes an energy crisis.
3. What are some signs of the shortage of energy?
4. What is a cartel?
5. Why is the OPEC group so powerful?
6. What helped cause severe inflation in the United States in 1974?
7. What effects might the energy crisis have in developed countries?
8. Why would underdeveloped countries suffer the most from an energy crisis?

9. Describe how the energy crisis has rearranged the wealth of the world.
10. Explain how you would solve the energy crisis.
11. What examples of energy waste can you find in your way of life?

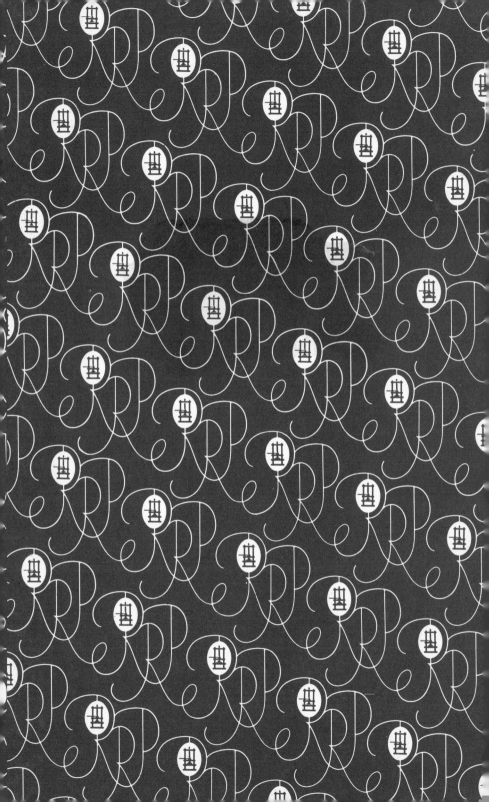